Routledge Revivals

A Rural Policy for the EEC?

First published in 1984, Hugh Clout's work contributes to one of the most debated and important topics of the time, the European Economic Community and its rural policy. Starting from the Mid-20th century, Clout explains the profound socio-economic and environmental changes that affected the countryside of Western Europe. This work shows how the EEC's wide-ranging Common Agricultural Policy added a measure of uniformity to farm policies. Clout reveals that the transformation however was not an entirely healthy one. The broad process of agricultural modernisation reinforced the numerical decline of farm workers throughout Western Europe, weakened many rural communities, and served to accentuate depopulation. Clout's work ultimately argues forcibly that to produce such a programme for managing rural Europe would be a major challenge for the EEC in the future.

T0298609

A Rural Policy for the EEC?

Hugh Clout

Routledge
Taylor & Francis Group

First published in 1984
by Methuen

This edition first published in 2018 by Routledge
2 Park Square, Milton Park, Abingdon, Oxon, OX14 4RN
and by Routledge
711 Third Avenue, New York, NY 10017

Routledge is an imprint of the Taylor & Francis Group, an informa business

© 1984 Hugh Clout

Publisher's Note
The publisher has gone to great lengths to ensure the quality of this reprint but
points out that some imperfections in the original copies may be apparent.

Disclaimer
The publisher has made every effort to trace copyright holders and welcomes
correspondence from those they have been unable to contact.
A Library of Congress record exists under the ISBN: 84006600

ISBN 13: 978-1-138-30775-9 (hbk)
ISBN 13: 978-1-315-14257-9 (ebk)
ISBN 13: 978-1-138-30778-0 (pbk)

A Rural Policy for the EEC?

HUGH CLOUT

METHUEN
LONDON AND NEW YORK

First published in 1984 by
Methuen & Co. Ltd
11 New Fetter Lane, London EC4P 4EE

Published in the USA by
Methuen & Co.
in association with Methuen, Inc.
733 Third Avenue, New York, NY 10017

Typeset by
Graphicraft Typesetters Limited, Hong Kong

British Library Cataloguing in Publication Data

Clout, Hugh
A rural policy for the EEC.—(The Methuen
EEC series)
1. Regional planning—European Economic
Community countries   2. European
Economic Community countries—Rural
conditions
I. Title
307′ .12′ 094   HT395.E82

ISBN 0-416-34540-9
ISBN 0-416-34550-6 Pbk

Library of Congress Cataloging in Publication Data

Clout, Hugh D.
A rural policy for the EEC?

(The Methuen EEC series)
English and French.
Bibliography: p.
Includes index.
1. Rural development—Government policy—European
Economic Community countries.   I. Title.   II. Title:
Rural policy for the E.E.C.?
HD1920.5.Z8C56   1984   338.94′ 009173′ 4   84-6600
ISBN 0-416-34540-9
ISBN 0-416-34550-6 (pbk.)

Contents

Abbreviations

CAP	Common Agricultural Policy
DATAR	Délégation à l'Aménagement du Territoire et à l'Action Régionale
EC	European Community(ies)
ECU	European Currency Unit
ERDF	European Regional Development Fund
FIDAR	Fonds Interministeriel de Développement et d'Aménagement Rural
IDP	Integrated Development Programme
IUCN	International Union for the Conservation of Nature and Natural Resources
OECD	Organization for Economic Co-operation and Development
PAR	Plan d'Aménagement Rural

Figures and tables

Figures

Tables

General editor's preface

The European Economic Community came into existence on 1 January 1958, having formally been established by the signature of the Treaty of Rome on 25 March 1957 and by its subsequent ratification by the governments of the original six member states. The Rome Treaty also established the European Atomic Energy Community (Euratom), and the European Coal and Steel Industry had been created in 1952 by the Treaty of Paris. These bodies, united since 1967 under a common Council and with a Common Commission and generally known as the EEC or the European Communities, are a powerful and complex force, affecting the lives of the citizens of all member states and the economies and policies of many non-member states.

Since 1958 many of the main policy objectives of the Rome (and the Paris) Treaty have been realized. There remain, however, policy areas where progress has been very slow and difficult, and on the whole it is these problems that draw attention and criticism. There is no doubt that the member states and sectoral interest groups of the enlarged and enlarging Community are still experiencing considerable difficulty in reaching an acceptable balance between national and Community interest, a situation that is not greatly assisted by the low level of general interest in the populace at large of the character, aims and procedures of the Community and its institutions.

The more widespread dissemination of information and opinion about the Community deserves higher priority than has hitherto been given. This series of books, in consequence, is designed to cater for the needs of both those with more specialist interest and those with a more general desire for ready access to fact and informed opinion. Each book is written by an expert on the particular subject, yet with a style and structure that will make it accessible to the non-specialist. The series is designed to facilitate the crossing of disciplinary boundaries and hence to encourage discussion and debate in a multi-disciplinary context (in the field of European Studies, for example) of one of the most powerful and dynamic communities in the world.

R. A. Butlin
Loughborough University of Technology

Author's preface

My interest in the changing countryside of Western Europe began when, as an undergraduate, I started to investigate detailed aspects of rural housing and land use in areas of Northern and Western France. In the following twenty years I have researched a number of rural themes in the recent historical geography of the French countryside and have studied current rural planning problems in the Massif Central and several other regions. My acquaintance with wider rural issues has been stimulated by travel to other parts of the European Community and by contacts with rural geographers who have been kind enough to share their research findings with me and often to talk about rural problems in the field. I extend my gratitude to them. In recent years I had the privilege of acting as rapporteur for a working group on social and economic problems encountered in rural parts of the EC, organized by the Institute for European Environmental Policy (IEEP) and involving such distinguished European politicians as Edgar Faure and Michel Cointat. I am most grateful to the IEEP for enabling me to perform this role, to obtain an international set of views of countryside matters, and to catch the occasional glimpse of the working of Europolitics in Brussels, Strasbourg and Paris.

The present book undoubtedly reflects my geographical orientation but I have tried to present themes in a wider interdisciplinary way with students pursuing courses in European Studies in mind. In particular I have drawn material from a broad range of publications, embracing agricultural economics, ecology, landscape planning and rural sociology as well as geography. Also I have included a number of items in French and other languages, knowing that many of these works will be accessible to students in European Studies. The preparation of this book owes much to the patience and skills of Alick Newman in the Cartographic Unit at University College London and of the secretarial staff in the Department of Geography and the Faculty of Arts. My warmest thanks are extended to these colleagues for their practical help in so many ways.

Hugh Clout
Bloomsbury

Acknowledgements

The author and publisher would like to thank the following for permission to include redrawn versions of illustrations:
Professor René Lebeau for Figure 1.3; Professor Jean Pitié for Figure 2.1; Professors André Fel and Guy Bouet for Figures 3.2, 5.7 and 8.2; Revue Géographique des Pyrénées et du Sud-Ouest and Professor Pierre-Yves Pechoux for Figure 4.2; and Dr H. Meijer and the Information and Documentation Centre for the Geography of the Netherlands, Utrecht, for Figures 6.5 and 6.6.

1 The Community's countryside

Prospect

In recent years there has been growing interest in the countryside of the European Community. Members of the general public in the UK have glanced nostalgically at country life through popular novels and a spate of television programmes derived from them. In France books like *Le Cheval d'Orgueil* (1975) by Pierre-Jakes Hélias have adopted a rather sharper view of rural conditions in the past, while handsomely illustrated countryside books and entrancing films on natural history have proliferated in every country of the Ten to indulge a desire by urban Europeans to look beyond their immediate surroundings. As well as seeing portrayals of the countryside in their own living rooms, more and more car-owning city dwellers travel into rural areas for rest and recreation and thereby sample them on a temporary basis. As ecology has come into fashion, numerous action groups have been founded to encourage the conservation of rural landscapes and the natural environment, and governments have responded by designating various types of cherished space (Lowe and Goyder 1983). Both public and politicians have mixed feelings about the Common Agricultural Policy, which all too often – and quite erroneously – tends to be equated with the full spectrum of rural management. Emotions frequently run high, with some accusing farmers of being the 'fat cats' of the EC or participating in what Marion Shoard (1980) has provocatively called 'the theft of the countryside'; and others pointing to examples of low incomes and sparse services in the countryside, only to be accused of indulging in 'the usual bleating' of the rural lobby 'which seems to think country folk should have all the advantages of rural splendour and city convenience' (Anon. 1982a, 25). Add to this the frequent wearing of 'rose-tinted spectacles' by urban visitors in search of 'the good old days', which so many advertisers tell them can be found in the countryside, and it is hardly surprising that some strange perceptions of rural life are encountered.

None the less these are tending to change as urban Europeans begin to realize that the countryside is not all calm and contentment. In many respects it is nothing short of disputed space, with increasing amounts being consumed by housing and other urban uses each year and profound disagreement encircling the precise uses to which the Community's very varied rural land should be put and the degree of modern technology that is desirable upon it (Wibberley 1981). Incomers to the countryside often espouse different views from those expressed by long-established country dwellers on the use of rural resources, and advocates of ecological principles clash with proponents of modern, intensive farming and rural 'development' in its many guises. Local residents often hold opinions about managing the countryside that contrast strikingly with those expressed by national governments and supranational organizations, while remarkably varied needs are felt and expressed by members of the different groups in society who happen to live in the countryside (Mormont 1983).

Faced with all these issues, it is rather surprising that so little has been written about land and people in rural areas of the EC as a whole. Detailed case studies abound, relating to specific topics at particular places, in individual regions or single countries, as do studies of the intricate workings, merits and disadvantages of the CAP, but when one looks out from farming and peers in the direction of the full range of rural matters in the EC the literature becomes distinctly sparse. However, many country dwellers, administrators and politicians in the member states are attempting to do precisely that and similar viewpoints are being adopted in supranational institutions such as the Council of Europe and the European Parliament. Ideas are emerging – and they are not really new at all – on the desirability of adopting ways of thinking about the countryside as a whole, rather than contemplating and planning agriculture, nature conservation, recreation and other aspects of life in isolation. Such people and organizations are not advocating a single inflexible 'common rural policy' which would be enforced across the Community's highly varied country areas but they are urging that an holistic view of problems be adopted, rather than the conventional compartmentalized one.

Without entering into scientific or technical detail or institutional complexity, the following chapters attempt to examine the evidence that lies behind this train of thought, stressing major processes that have given rise to changing lives and changing landscapes since mid-century, have produced numerous problems and clashes of interest, and will generate further transformations in the years to come. After discussing the fragile resource base of the EC and its diverse cultural landscapes, rural circumstances will be portrayed *c.* 1950 in order to provide a benchmark against which present conditions can be measured. Thereafter the approach will be predominantly systematic, although the internal diversity of rural Europe demands that national and regional examples be included to refine broad international generalizations. One overarching theme will be the ways in which changing pressures for economic production, residence, recreation and conservation have transformed the shrinking amount of space that may be described as countryside, while a second will be how rural areas have been planned to cope with these pressures in recent decades and how they might be managed in the future.

Context

With 270,967,000 people living on 1,657,621 km² the EC forms one of the most densely populated sections of the Earth's surface. It is also one of the most urbanized, with almost four-fifths of those inhabitants

Table 1.1   Area, population and urbanization, 1981

	Area ('000 km²)	Population ('000)	Density (per km²)	Urban land (%)	Urban population (%)
Belgium	30,519	9,859	323	15.5	95
Denmark	43,080	5,123	119	10.0	80
France	543,965	53,716	99	5.6	78
W. Germany	248,667	61,566	248	13.2	85
Greece	131,990	9,707	74	4.0	65
R. of Ireland	70,285	3,401	48	1.7	52
Italy	301,266	57,070	189	5.2	65
Luxembourg	2,586	365	141	7.5	68
Netherlands	41,160	14,150	344	16.3	88
UK	244,103	56,010	229	9.0	78
EC	1,657,621	270,967	163	7.8	77

Source: The source of all tables, unless specified otherwise, is the European Community's official Eurostat collection.

living in what the individual nations have recognized as 'urban' settlements and the Ten containing a score of cities with over 1,000,000 residents apiece (Table 1.1). Large sections of land are given over to urban and industrial uses, with megalopolitan regions (having more than 250 or even 500 inhabitants/km²) sprawling over vast areas and additional stretches being built on every day. But such features make up only part of the picture, for although national densities surpass 300/km² in the Netherlands (344) and Belgium (323) and average out at 163/km² for the EC as a whole, Western Europe contains large relatively empty areas, with France (99), Greece (74) and the Republic of Ireland (48) recording low national densities and values falling below 25/km² in areas of upland Britain, many sections of interior France, the Alps, Corsica, Sardinia and parts of Greece (Béteille 1981) (Figure 1.1).

Despite the powerful march of urbanization in recent decades only 6.8 per cent of the surface of the EC was devoted to urban land uses in 1971 (Best 1979, 1981). Fully comprehensive figures have not been collated for later dates but if one pieces together greatly varying evidence from statistical yearbooks and makes some judicious inter-polations on the basis of trends during the 1960s, it would appear that in 1981 roughly 7.8 per cent of the surface of the EC was devoted to urban uses, embracing housing, factories, roads and other transport uses (Table 1.1). On the basis of this calculation, over one-eighth of all land is now used for urban purposes in the Netherlands, Belgium and West Germany, while the figure is not much over 5 per cent in France and Italy and is even lower in Greece and the Republic of Ireland. The remaining nine-tenths of the EC is devoted to agricultural activities, forestry and conservation and forms the mosaic of 'rural space' or, more simply, the 'countryside'.

In the 1980s the EC is emphatically both 'urban' and 'rural' at the same time. Most of its people inhabit towns and cities and have lifestyles very removed from cultivating the soil, tending livestock or producing timber; but by far the greater share of its land is used for these very activities. 'Town' and 'country' were long thought of as contrasting states although, of course, there were important functional links between them, since town dwellers drew their food supplies from the countryside, often invested in the ownership of rural land, and dispatched commodities back there (Kayser 1972). In recent years these concepts have become blurred in many complicated ways in response to profound changes in people's working and living habits (Juillard

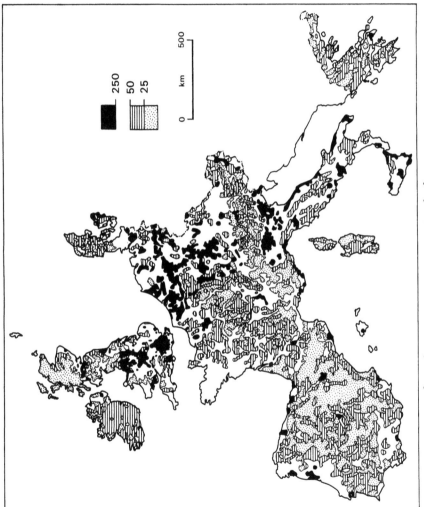

Figure 1.1 Density of population in Western Europe (per km²)

1973). By virtue of newspapers, radio, television and personal contacts with city dwellers, rural residents are well aware of urban lifestyles, norms and values. Similarly the productive activities of most farmers are geared to satisfying urban demands for food supply. Provision of public and private transport has allowed many rural dwellers to retain possession of their home or family farm and yet travel to urban employment on a daily or weekly basis (Mathieu and Bontron 1973). In many such instances their rural property is run as a free-time agricultural holding. Car-owning city dwellers in their millions have chosen to live in spacious suburbia or to move even further out into commuter settlements, which retain something of their village-like appearance, even though their socio-economic characteristics have changed dramatically. Shorter working weeks (and years), combined with a greater proportion of disposable income for non-essential items, and high levels of car ownership, have fostered the temporary presence of large numbers of city people in the countryside, associated with all forms of rural tourism through to the acquisition of second homes (Kayser 1973). The growing trend for retirement migration brings large numbers of elderly people into certain rural areas of the EC, thereby helping to buoy up their residential totals but giving rise to new sets of problems. Finally, the last two decades have seen the emergence of a trend for urban dwellers to 'drop out' of their established routine, resign their job (if they have one), and aspire to enjoying the good life of a small country living.

Varying combinations of these processes are at work in the many expressions of rural space which make up the Community's country-side. Town and country are certainly not as clearly distinguishable as they were in the past and some would argue that from a social point of view labels like 'urban' and 'rural' are unhelpful and downright confusing, since all West Europeans are now exposed to the ideas and attitudes of the city even if they do not live *in* a city (Pahl 1965, 1975). City and countryside have become integral parts of the same socio-spatial system, thus changes in either component are interdependent. Doubtless that is correct and one would be disappointed if one were to think it would still be possible to discover the kinds of relatively enclosed, inward-looking, farm-dominated and partly self-sufficient rural community that existed in the past (Canevet 1979). A whole new vocabulary has been created to describe the current situation in which the countryside is part of a wider, open system (Hanrahan and Cloke 1983). After suburbia, the EC has its share of 'exurbia', 'dispersed cities', 'metropolitan villages' and territory where the processes of 'rurbanization' are at work (Bauer and Roux 1976). Some would argue that 'counterurbanization' and the emergence of 'regional cities' are now the dominant processes of spatial transformation in the EC, as inner cities and old-established suburbs lose population but surrounding villages and small towns grow prodigiously (Bryant, Russwurm and McLellan 1983). One must, of course, pay full recognition to the impact of these influences on the countryside but it is impossible to escape the fact that farms, fields and forests *are* materially different from factories and city streets, and that quantities of rural space are being converted to urban uses every year (Berger and Fruit 1980). In visual, formal and functional terms, countrysides *do* exist and it is with rural space and the people who live and work in it and visit it that this book is concerned.

The notion of 'countryside' may once have appeared to be a fundamental and simple concept but it is actually very difficult to define in precise terms (Bonnamour 1973, Enyedi 1975). It certainly does not correspond automatically with administrative areas that are labelled officially as 'rural' nor may it be defined by any single criterion of population density or settlement arrangement (Mendras 1980). For example, parts of Southern Europe contain settlements of up to 50,000 people that are strongly concerned with agricultural activities and function as veritable 'agrotowns' (King and Strachan 1978). None the less, for operational purposes, it may be said that rural space comprises areas with:

relatively low densities of population, including localities with small towns of up to 20,000 people (and, indeed, larger 'agrotowns' in Southern Europe);

relatively loose networks of infrastructure and services;

relatively tight networks of personal contact and/or identity with the locality;

below-average proportions of manufacturing and office-based activities;

and a dominance of farmland and/or forestry in the land-use mosaic (Anon. 1980a).

Stretches of countryside have been fashioned by numerous generations over many centuries (indeed, millennia) and continue to be remodelled with every year that passes. Their diversity and intricacy is one of the distinguishing characteristics of Western Europe, as compared with other parts of the world of similar size, and it is appropriate to examine that legacy and the fragile resource base from which it has been created before identifying the profound changes that have occurred in rural Europe since the middle of the twentieth century.

Resources

The EC occupies a broad swathe of environmental conditions extending across 25° of longitude and through 35° of latitude from Northern Scotland to the Mediterranean milieux of Greece, and from the Atlantic fringes of the British Isles to the continental interior of Bavaria. Juxtaposition of land and sea, in the form of so many peninsulas and islands, and the presence of suites of plains, hills and high mountains introduce further dimensions of diversity to the land base of the Ten. Large areas of uniform topography are relatively rare. Each valley floor and hillside, indeed each facet of land, differs from all others in details of soil composition, microclimate and vegetation potential but it is possible to classify Western Europe into Atlantic, Subatlantic (or transitional), Central European (mainly Alpine) and Mediterranean phytogeographical zones or domains (Ozenda 1979) (Figure 1.2). Without human interference, deciduous broadleaf forest would occupy much of the first two domains, where the subsequent accumulation of leaf mould would serve to enhance soil quality. Mixed woodlands of evergreen and deciduous trees would be found in parts of central Europe, while coastal areas of the Mediterranean zone would be covered by open woodland of broadleaf evergreen trees, interspersed with woody shrubs and grasses, all being able to withstand conditions of summer drought. Localized variations in altitude, humidity and soil quality would introduce more diversity, clothing Central Europe's mountain slopes with evergreen trees and high peaks with natural grassland, supporting marsh vegetation in damp valleys and low heathland plants on stretches of poor acid soil.

The natural basis of Western Europe's countryside would give rise to a relatively straightforward vegetation mosaic but this has been modified in myriad ways over the millennia of human occupation and continues to be transformed in drastic fashion in the second half of the twentieth century. Great stretches of woodland have been cleared to provide space for cultivation and for grazing livestock, while timber has been felled to provide fuel and raw material for house construction,

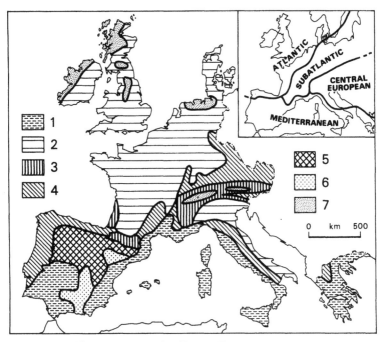

Figure 1.2 Vegetation zones in Western Europe
1 Mediterranean open forest
2 Broadleaf deciduous forest
3 Evergreen forest
4 Mixed coniferous and deciduous forest
5 Mixed broadleaf evergreen and deciduous forest
6 Grassland
7 Heath, moor or Alpine vegetation

shipbuilding and numerous other crafts. Rural settlements, field boundaries and other visible components of the landscape have been installed across the face of rural Europe as a result of centuries of human endeavour. Stones have been removed painstakingly from cultivated soils and water levels have been controlled carefully. By far the majority of marshes have been drained, so that Western Europe's remaining wetlands form particularly rare and vulnerable environments of great interest for the vegetation and wildlife they support. Similarly, most expanses of heather, gorse or broom, which occupied areas of poor acid soils, have disappeared as a result of human activity. In centuries past they clothed parts of the sandy plains of

Aquitaine, the Sologne and northern Germany and also covered soils derived from sandstones, schists and granites along the Atlantic fringes of Britain and Ireland and in Armorica. Their traditional functions included periodic cultivation and sheep grazing but during the last hundred years most of these areas have been reclaimed for cultivation or planted with conifers. Those that remain are fragile and unusually attractive ecological survivors.

Centuries of agricultural occupation in Atlantic and Subatlantic Europe certainly transformed soil profiles and humanized the vegetation in very profound ways, indeed most of the woodland to be found there is the result of reafforestation. This has produced a radical change in tree species, with softwoods such as pine, spruce and larch being planted in vast quantities in preference to traditional hardwoods (Noirfalise 1979). For example, it has been estimated that in West Germany reafforestation over the past two centuries has reversed the former ratio of one-third conifers to two-thirds broadleaved deciduous trees, as multipurpose woodlands, which yielded grazing space and a range of commodities as well as wood, were replaced by viable productive stands of timber (Scheifele 1979). Beyond the woodlands, complex techniques of working the land, raising livestock and restoring soil fertility were gradually devised to enable rich yields of agricultural produce year after year, generation after generation. The natural resource base was, of course, changed profoundly but arguably was transformed into a more worthwhile set of exploitative systems which normally maintained the soil in good heart. Farming was essentially a manual operation, requiring large inputs of labour and supporting great numbers of countryfolk. As a general rule, agricultural practices were conducted with due respect for natural conditions of soil, climate, slope and season, although there were notable exceptions. Modern agricultural systems have given farmers great freedom from their traditional constraints and an unparalleled ability to do both good and harm in the rural environment (Poore and Ambroes 1980). Deep or wide areas of wetland, which had always been considered unreclaimable, have been drained and cropped in recent years; numerous stretches of moor and small patches of timber have been cleared; heavy farm machinery has been used on fragile soils; and ever greater quantities of artificial fertilizer, herbicide and pesticide are being used on farmland and are giving rise to serious environmental problems, including a substantial decline in wildlife.

The main exception to the view that traditional exploitation systems incorporated wise husbandry of resources is to be found in the rural history of Southern Europe. Comparing the testimony of classical writers with present conditions shows that the ecosystems of Mediterranean Europe have proved particularly susceptible to deterioration. Thus, Plato described abundant forests in the mountains of Attica, where many livestock were grazed and water could be stored to supply springs and rivers (Pelapsis and Thompson 1960). In our times much of that part of Greece and, indeed, many other sections of the Mediterranean world have become arid and barren. Throughout Southern Europe large areas of open forest, comprising broadleaved evergreen hardwoods which were well adapted to long dry summers, have been destroyed across the centuries through a combination of felling, overgrazing by sheep and goats, and because of natural and man-induced fire (Thirgood 1981, Naveh 1982). Level, potentially cultivable land is scarce throughout the islands and peninsulas of the Mediterranean. For example, only a quarter of the total cropland of Greece possesses those qualities and most of that is in isolated mountain valleys and discontinuous stretches along the coast. Unfortunately, much of the lowland is in the arid south-east of the country, where climatic conditions are least favourable for cultivation. In past times the shortage of suitable low-lying terrain in Greece, Italy and Mediterranean France made it necessary for steeply sloping land to be cultivated and in many locations cropping was extended beyond wise ecological limits, with open arable fields being particularly subject to erosion. Scrupulous creation of terraces could control harmful effects but where such work was undertaken carelessly or was omitted altogether disaster could easily ensue.

Rainfall in Southern Europe is concentrated seasonally, being particularly heavy during spring and autumn when there is little plant cover on cultivated or devastated soils. In autumn the rain falls on parched top soil and washes over readily. As well as being composed of sloping land extensive Mediterranean areas display surface deposits that are rich in silt and clay and have been carved into deep gulleys or have been stripped away to reveal underlying materials. Areas of bare rock are the most extreme result of such exploitation but more usual are stretches of evergreen shrubs which form a low irregular ground cover. Rivers and streams experience great seasonal variation in flow and become laden with gravel, sand and silt that will be deposited in lowlying areas. Deposition of topsoil can improve land downslope but any beneficial effects are soon lost once such areas are overlaid with eroded subsoil, fragments of underlying rock and other infertile materials. In addition, lowlying areas may be subject to aeolian erosion since strong winds can easily erode parched deposits in summer and give rise to highly immature soils. In short, the natural resource base of Western Europe has been transformed into a series of ever-changing cultural landscapes of enormous internal diversity and intricacy and of great fascination.

Landscapes

The landscapes of the EC are like a set of palimpsests that many settlers have inscribed after erasing in part or entirety the marks that had been placed there by earlier occupants. The characteristics and the changing appearance of these cultural landscapes have been investigated by several generations of anthropologists, ecologists, geographers and historians, with members of each discipline emphasizing different aspects of the rural environment, including agricultural technologies, building materials, rural house styles, field arrangements and settlement types. Space forbids consideration of the intricacies of their findings, indeed all that will be offered is a brief sketch of three aspects of human occupation in the countryside, namely rural patterns, traditional construction materials and building styles.

According to Lebeau (1969), Western Europe in the second half of the twentieth century may be characterized by three main types of rural pattern, namely openfields, enclosures and Mediterranean landscapes, that are identified on Figure 1.3, while morphologies of rural settlement are shown on Figure 1.4. Predominantly arable open landscapes with irregular nucleated settlements remain widespread across north-eastern France, Wallonia and Central and Southern Germany. By contrast, they were swept away in the past in lowland England and Denmark and were replaced by enclosed fields, bounded by quick-set hedgerows. Their settlement patterns were enhanced by establishing hamlets in parts of lowland England and scattered farms also characterize the more pastoral farmlands of Ireland, upland Britain, Southern and Western France, Alpine areas, Northern Germany and parts of the Low Countries. The latter areas are fringed by distinctive rural landscapes whose regularity of field pattern and linear village morphology derives from planned reclamation of marsh. Similar features in parts of Central Germany are the product of systematic clearance of woodland for cultivation. Finally, there are the

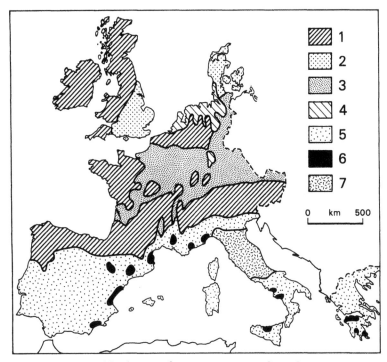

Figure 1.3   Rural landscapes of Western Europe (after Lebeau 1969)
1 Enclosed landscapes, dispersed settlement, much permanent grass
2 Former openfields, grouped settlement, subsequent enclosure and settlement dispersion
3 Openfield, grouped settlement, much arable
4 Linear settlements, polderland or forest
5 Mediterranean openfields, tree crops, mainly grouped settlement with some dispersion
6 Huertas
7 *Coltura promiscua*

varied rural landscapes that have been fashioned from the fragile environments and contrasting terrain of Mediterranean Europe. Traditional settlements are normally nucleated and fields tend to be open and devoted to cereal cultivation, with some patches of olives, vines and tree crops. In Northern Italy settlement is more scattered and in some other areas dispersed farmsteads have been installed in the countryside between 'agrotowns' and large villages in recent years. Much more intensive systems of cultivation and higher densities of

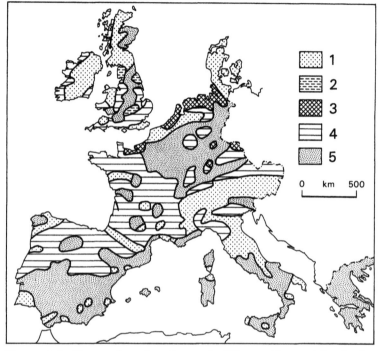

Figure 1.4 Forms of rural settlement in Western Europe (after Jordan 1973)
1 Scattered farmsteads
2 Green villages
3 Linear settlements
4 Hamlets
5 Irregular clustered villages

settlement are possible in irrigated areas throughout the Mediterranean world (Gade 1978). Another form of intensively worked landscape is found in areas of *coltura promiscua*, where tree crops and vines alternate with bands of cereals, fruit bushes and vegetables on valley floors and lowlands, with surrounding hillslopes being terraced carefully. The best-known examples are in Tuscany, Umbria, Emilia and Campania but *coltura promiscua* is also to be seen in Southern France, Greece and Iberia.

Within such complex arrangements of fields and settlements Western Europe displays a rich heritage of vernacular building styles fashioned from a variety of materials, since the link between rural building and the local physical resource base was much closer in the past than now (Figure 1.5). Use of stone was dominant in much of Mediterranean Europe and in many regions of France and the British Isles where deforestation had been particularly thorough and where suitable rock for building was quite easily accessible. Timber construction was characteristic in the Alps and the Central Pyrenees, and the practice of filling a timber frame with a woven wattle of twigs and clay ('half timbering') was widespread in Germany, Denmark and parts of England (Jordan 1973). Mud bricks were used for making walls in a number of Mediterranean lowlands where stone was not

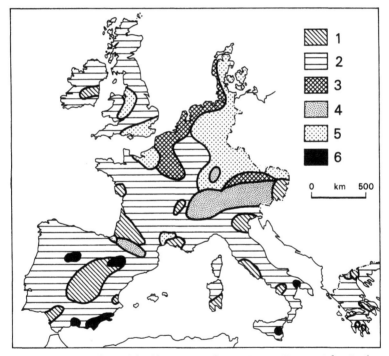

Figure 1.5   Traditional building materials in Western Europe (after Jordan 1973)
1 Sod, turf or clay walls
2 Stone
3 Brick
4 Wood
5 Half timbering
6 Areas with cave dwellings

available and cut turf was used in a similar fashion in Central Ireland. Techniques were more sophisticated in the North Sea lands, from Northern France through the Low Countries to Jutland, and in Bavaria where bricks were baked from local clays (Christians 1982).

The traditional farm buildings that resulted from the use of these local materials were constructed according to variants on two basic models: unit farmsteads (where accommodation for animals and humans as well as storage space was provided in a single building) and multiple-building farmsteads (where several separate buildings served these functions). Single-storey unit farmsteads were traditional in Northern Germany and the Low Countries, while several-storey unit farms were common in southern Europe and the Alps, affording space for livestock at ground level, with human accommodation on the first floor and storage space higher up. Multiple-building farms were arranged to form enclosed farmyards in Denmark, Central Germany and north-eastern France, while farmyards were not entirely enclosed by buildings in north-western France. Finally, traditional farm buildings were often clustered in more irregular fashion in the British Isles. Construction of main roads and railways after the mid-nineteenth century enabled building materials to be transported from distant areas and assisted the diffusion of new construction styles. For example, in Western France new types of building penetrated the environs of Le Mans and Rennes at much earlier dates than in Southern Normandy which remained isolated for much longer (Flatrès-Mury 1970). Likewise, innovations in agricultural activity in commercially orientated districts generated important modifications to old farm buildings.

These legacies of traditional rural Europe were, of course, created under varying conditions of social and economic control and at different stages in the past, with some dating back several centuries and others being much more recent (Santos 1978). Phases of population growth witnessed cultivation being extended from lowlands on to less suitable hillsides, slopes being terraced painstakingly, marshes and moorlands being reclaimed, and additional permanent settlements being installed. Periods of population loss were characterized by the retreat of the 'frontier' of cultivation and the shrinkage of settlements and surrounding farmlands. Changes in the balance of political power or economic circumstances could cause one cultural landscape to be replaced by another, sometimes with great speed and thoroughness (as in the case of the English enclosures) but more often in a lingering, piecemeal erosive fashion (Lacoste 1977). In short, the surviving remnants of traditional Europe, which still made up a significant proportion of the EC's countrysides after the Second World War, were by no means 'original' but they did share certain characteristics. They were fashioned as part of a predominantly hand-made world at times when rural labour was much more abundant and modernization was either unknown or exceptionally rare; and they related to conditions of production in which local systems of interaction were overwhelmingly predominant. By the middle of the twentieth century they were poised to experience human and environmental changes of greater pace and magnitude and more dramatic form than ever before.

2 Rural Europe at mid-century

Land and people

At mid-century the Second World War had been over for five years, all ten countries were starting along the difficult path of recovery and each in its own way was about to embark on a quarter-century of remarkable economic growth that would have profound implications on lifestyles and land uses and give rise to new systems of town-country relations. Much of traditional rural life could still be found in continuity with times past but notable examples of drastic change were in evidence which heralded the kind of transformation that was to become commonplace in subsequent decades. In 1950 the Ten contained a total population of 222,400,000 that already included two dozen metropolitan areas with more than 1,000,000 residents and incorporated nine individual 'millionaire' cities (Kosiński 1970). Indeed eight had already achieved that status on the eve of the war. At mid-century the metropolitan areas and more tightly defined 'cities' of London (10,491,000 and 3,273,000 respectively) and Paris (6,763,000 and 2,850,000) stood out clearly at the top of the league.

Official definitions of 'urban' settlements varied considerably, being related in some countries to thresholds of population concentration while merely reflecting administrative status in others, but three-fifths of all West Europeans were already living in towns and cities, with an urbanized 'core' contrasting with much more rural peripheral areas (Table 2.1). Just over 70 per cent of Dutch and West Germans were town dwellers but the UK had no less than 78.5 per cent of its population living in urban settlements. Indeed as far back as 1851 half of the population of England and Wales had been recorded as town-dwelling but one hundred years later that proportion still had not been reached in Greece, Italy and the Republic of Ireland and had only just been passed in France (54 per cent). Viewing the situation another

Table 2.1 Urban population and tenure of agricultural land, 1950 (percentage of area)

	Urban % of population	Farmland as % of total land	Owner-occupation	Tenancy	Share-cropping	Woodland as % of total land
Belgium	63.4	47.4	32.3	67.7	0.0	19.9
Denmark	67.3	67.4	93.0	7.0	0.0	10.3
France	54.1	58.5	52.0	41.5	6.5	20.7
W. Germany	72.5	49.3	85.3	14.7	0.0	28.1
Greece	37.0	70.0	92.4	5.5	2.1	14.9
R. of Ireland	40.6	81.1	93.8	6.2	0.0	2.0
Italy	44.0	59.3	69.6	14.6	15.8	19.2
Luxembourg	58.8	50.0	73.6	26.4	0.0	32.9
Netherlands	70.5	49.2	47.5	52.5	0.0	7.6
UK	78.5	77.5	55.0	45.0	0.0	6.8
EC	60.0	61.5	66.8	27.7	5.5	17.5

way, 45,000,000 lived in administrative areas that were deemed to be 'rural' but that figure was undoubtedly an understatement since many so-called urban settlements were very small, with nucleations of only 2000 inhabitants (and sometimes less) forming the critical threshold for urban status in several countries.

By contrast with these residential indices of urbanization, relatively little of the land surface of the future EC was devoted to housing, factories, railways, roads, airfields and intercalated open spaces (e.g. sports grounds, parks and cemeteries). Accurate evidence on land use is notoriously difficult to obtain; however painstaking work by Best (1979, 1981) showed that only 6.0 per cent of the EC was devoted to urban functions in 1961, at a stage well into the post-war phase of demographic and economic growth. Five per cent might be a reasonable figure for mid-century. There must have been significant national deviations from that norm, with well above average proportions of land in urban use in Belgium, the Netherlands and West Germany. Rather surprisingly, 95 per cent of the land surface of the long-urbanized countries of Western Europe was still devoted to rural uses, comprising cropland, pasture, woodland and the tantalizing category of 'other' miscellaneous uses (including military training land, open-cast mining and barren mountain areas). Roughly 17.5 per cent of the

Ten was under varying qualities of woodland, with West Germany (28 per cent) and tiny Luxembourg (33 per cent) being particularly well endowed but just the reverse being the case in the Republic of Ireland (2.0 per cent), the UK (6.8 per cent) and the Netherlands (7.6 per cent). When due allowance is made for towns and trees, 77.5 per cent of the Ten was devoted to the full range of farming activities in 1950. Despite the massive rise of interwar urbanization, it is arguable that Patrick Abercrombie's words, pronounced a quarter of a century earlier and with regard to just one country, held a great deal of truth for Western Europe as a whole: 'the most essential thing which is England, is the countryside, the market town, the village, the hedgerow trees, the lanes, the copses, the streams and the farmsteads' (cited by Lowe and Goyder 1983, 18).

Information on employment was not much more reliable than that on residence and land use but it seems that almost 24,000,000 people were employed (and in some cases underemployed) in agriculture in 1950, with particularly large contingents of farmers and farmworkers in Italy (almost 8,300,000), France (5,400,000) and West Germany (5,000,000), and the total farm workforce in the remaining seven countries only amounting to a further 5,000,000. Farming accounted for just under 30 per cent of the Ten's workforce but displayed great national and regional variation around that mean, falling to 6 per cent in the UK but rising to half the workforce in the Republic of Ireland (47 per cent) and Greece (52 per cent). For Italy as a whole 41 per cent of the workforce was in agricultural employment but the figure rose to 55 per cent for all of the Mezzogiorno and exceeded 70 per cent in some southern provinces (Barberis 1968).

High proportions of the workforce in agriculture were positively related to rural poverty. In parts of Greece and Southern Italy prevalent conditions were not far removed from subsistence, save in localities with market-orientated farming such as the production of fruit and vegetables in the Campania and parts of Sicily. Many peasant farmers in Mediterranean Europe continued the centuries-old struggle to wrest a living from their plots of fairly unproductive soil and in many areas the volume of labour was too great for the quantity of work available, especially during slack seasons (McEntire and Agostini 1970). For example, at mid-century Italy contained 3,500,000 farmers (owner-occupiers and tenants), 2,000,000 'active' family members who worked the land, and no fewer than 2,700,000 landless labourers. Many of the latter obtained low-paid farmwork for half of the year if they were fortunate and in many cases for only 80–120 days. Large proportions of all three groups were enmeshed in a syndrome of perennial poverty, embracing not only poor and insufficient food, bad housing and material deprivation, but also inadequate education and low social status, since they were all subordinate to larger landowners and members of the middle class in the villages.

Rural poverty was less extreme beyond Mediterranean Europe but most farmers and labourers in almost all countries of the Ten continued to live frugally, much as they had always done. Electrification was far from complete in the countryside at mid-century and was particularly deficient in the rural fringes of Southern and Western Europe, but it brought some comfort to poor dwellings in many

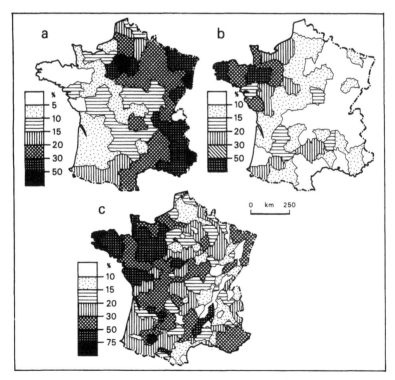

Figure 2.1 Rural housing quality in France *c.* 1950 (after Pitié 1969)
a With running water (per cent)
b Without electricity supply (per cent)
c Lacking all specified amenities (per cent)

regions that contained precious little furniture or any modern equipment. As a result, 'even the most modest working class home in a town would seem luxurious to a peasant's eyes' (George 1949, 119). Travellers in many parts of France, and especially Brittany and the Massif Central, were struck by evidence of 'profound poverty and the remarkable delapidation of agricultural buildings' since many farmhouses dated back 100 or even 150 years (Boichard 1958, 26). Comparisons with the UK and other neighbouring countries of north-west Europe were nothing short of 'humiliating' (Leroy 1960, 7). Western France was particularly deficient in piped water supplies and Brittany lagged behind the rest of the country in terms of electrification (Figure 2.1). Most farmhouses in the cool north-west lacked any means of domestic heating, save the kitchen range (Pitié 1969). Many rural dwellers in Western Europe were living lives at mid-century which differed little in material terms from those of their grandparents.

Landholding

There were almost 10,000,000 farms of more than 1 ha apiece in the Ten in 1950, of which Italy and France together accounted for virtually three-fifths. In addition, there were numerous tiny holdings which fell beneath that size threshold but were not recorded consistently. Over 85 per cent of all recorded farms in Belgium and Greece were between 1 and 5 ha in size and exceeded 55 per cent in West Germany and the Netherlands (Dewhurst, Coppock and Yates 1961). Unfortunately information that is directly comparable is not available for Italy but it would simply have confirmed the general impression. There is no doubt that the bulk of Western Europe comprised a mosaic of small farms which depended on family labour for their operation. In contrast, large capitalistic estates, which relied on hired labour, were exceptions to the rule but they existed in modern form in lowland England, the Paris Basin and some other regions, and much more traditional versions were encountered in Italy where they were soon to be the object of land reform. A second distinction may be drawn between peasant and commercial systems of farming. A majority of farmers on the agricultural periphery of Southern and Atlantic Europe still produced most of their goods for consumption by their own family and whatever surpluses were available were only dispatched to the nearest market. For these peasants agriculture was literally a way of life (Franklin 1969, 1971). However, orientation toward market requirements was strongly developed in Denmark, England, Northern France and some parts of the Benelux countries, Northern Italy and West Germany.

Yet another contrast existed with respect to land tenure which in Western Europe was spared the radical changes that were under way in the countries of Eastern Europe (Medici 1969). Two-thirds of Western Europe's farmland was worked by its owners in the early 1950s and this represented almost the only formula in Denmark, Greece and the Republic of Ireland, where more than 90 per cent of farmland was owner-occupied, and in West Germany, where 85 per cent was operated in this way (Table 2.1). Exploitative landlordism had brought tenancy into ill repute in some parts of Western Europe and measures had been taken in Denmark and the Republic of Ireland to emancipate small farmers from its evils. Landlord-tenant relationships remained virtually feudal in the Mezzogiorno and were to continue that way until well into the post-war period. Efficient, non-exploitative forms of tenancy worked well in the UK, the Benelux countries and Northern France, where they embraced substantial proportions of the farming surface. Share-tenancy, whereby 'rents' were paid in kind rather than cash, accounted for less than 6 per cent of Western Europe's farmland at mid-century and was being abandoned in favour of normal tenancy, although it still involved one-sixth of Italy's agricultural surface. In reality, the situation was not quite so straightforward since many functioning farms were not exclusively owner-occupied or tenanted but instead incorporated complex assemblages of fields and plots that were held under several tenurial regimes.

As well as the division of Western Europe's farmland into just under 10,000,000 recorded farms exceeding 1 ha apiece, many holdings were pulverized into a number of tiny plots or strips that were scattered over considerable areas and were intermixed with the plots of other owners. This kind of fragmentation derived from several origins. Some of it represented a fossilization of openfield patterns produced by historic systems of communal farming, but other examples resulted from the operation of laws or customs relating to inheritance that required an equal division of property between heirs. The gavelkind system which operated in much of Central and Southern Germany served to produce such an effect, as did the Napoleonic Code of land inheritance in France. Because of the splitting up of properties, villages became more and more crowded and plots became smaller, since they were not shared out between heirs but each single one was divided. In other regions of the Ten, farms were transmitted intact from one generation to the next and this kind of extreme fragmentation was avoided. Thus in Northern Germany and in the Black Forest both law and custom stipulated that farms should not be divided and in other parts of the country it was the tradition for each holding to be handed on to a single successor (Sperling 1966). In the UK, Denmark and some parts of Germany enclosure had consolidated once-fragmented patterns of landholding in earlier centuries, but in France, Greece, the Benelux countries and many regions of Germany and Italy this kind of reorganization had not happened and property fragmentation remained a major aspect of the countryside. Indeed, rural planners estimated that almost half of all Western Europe's agricultural land was so seriously fragmented as to merit consolidation, with 50 per cent of farmland in West Germany and 40 per cent in France needing urgent attention but the figure falling to a mere 5 per cent in Denmark.

Farm production

From the preceding discussion of employment and tenurial relations it is clear that levels of agricultural efficiency varied enormously within the Ten, with the north-western countries and the Mediterranean lands occupying extreme positions on the spectrum. Over the greater part of Western Europe farm mechanization was at an early stage in 1950 and heavy reliance was still placed on human labour and the use of horses and other farm animals. Only 725,000 tractors were in use and over two-fifths of these were in the UK, where the ratio between tractors and farms of more than 1 ha was 1:1.5 (Table 2.2). This stood in contrast with Western Europe as a whole (1:12) and was brutally different from conditions in Italy (1:62) and Greece (1:100). Indeed less than 2 per cent of Greek farms used any kind of mechanized power throughout the 1950s and over a quarter had neither machinery nor animal power, relying entirely on human muscle (Pelapsis and Thompson 1960). In 1950 farm technology in most parts of the Ten was still essentially traditional and manual in character, but conditions were to change rapidly and many farmers in the northern countries were soon to purchase tractors. By 1955 the total for the Ten was to rise to 1,340,000, of which half were to be in the UK (421,000) and West Germany (384,000). The tractor/farm ratio was to reach 1:1.1 in the UK, to be followed by 1:4 in both Denmark and West Germany,

Table 2.2   Aspects of agricultural mechanization and productivity *c.* 1950

	Fertilizers kg/ha farmland⁺ 1950	Tractor/ farm ratio 1950*	Tractor/ farm ratio 1955*	Combine/ farm ratio 1955*	Wheat yields 100 kg/ha 1948-52	Potato yields 100 kg/ha 1948-52
Belgium	169	28	12	264	31.1	235
Denmark	83	12	4	120	35.0	176
France	40	15	8	140	17.8	131
W. Germany	105	12	4	278	25.8	244
Greece	13	100	95	n.a.	11.0	n.a.
R. of Ireland	19	24	12	266	22.5	214
Italy	24	62	28	5,682	16.5	63
Luxembourg	n.a.	n.a.	4	331	18.2	185
Netherlands	189	12	5	127	32.3	240
UK	70	1.5	1.1	17	26.4	193
EC	n.a.	12	6.5	160	18.7	182

Notes: *Expressed as 1 : *x*; and normally rounded to the nearest whole number
⁺Excluding rough grazing.

but the average for the Ten was only 1:6.5. Not surprisingly, a similar north/south contrast was to apply to the adoption of combine harvesters, of which 27,700 out of a total of 56,000 were in the UK where the combine/farm ratio was 1:17 in 1955 in contrast with 1:160 for the Ten as a whole.

Much of the disruption caused to farm production during wartime had been overcome at mid-century and most countries were well on the path to recovery. The Ten produced 23,100,000 metric tons of wheat, of which one-third came from France and Italy where particularly large surfaces were devoted to this crop. Barley output was much smaller (7,450,000 metric tons) and was dominated by northerly countries (UK, Denmark, France and West Germany), while West Germany produced three-fifths of the total output of rye (4,900,000 metric tons) and half of the potato crop (66,500,000 metric tons). Indeed, potatoes continued to provide an essential component in the staple diet of many West Europeans whose meals were regulated by the dictates of food rationing. Patterns of productivity confirmed the by now familiar north/south contrast. On average each hectare of land sown with wheat in the Ten at mid-century produced 18,700 kg but national productivity varied from strikingly high yields in Denmark, the Netherlands and Belgium to the modest output of Italy and Greece (Table 2.2). Similarly, potato yields were high in the northerly countries of the Ten, with West Germany, the Netherlands, Belgium and the Republic of Ireland far above the average (182,000 kg).

The underlying reasons for these and other variations in productivity were bound up with variations in the physical resource base and differing histories of land occupation and market orientation, but a more immediate cause related to striking differences in the volume of fertilizer being applied to agricultural land. Large, and in some instances exceptionally large, quantities were already being used in the Netherlands, Belgium, West Germany and Denmark, by contrast with very modest use in the rural fringe of Southern Europe and the Republic of Ireland (Table 2.2). Livestock husbandry took more time to recover from the disruption of war but by mid-century the Ten supported 57,872,000 head of cattle, 49,708,000 sheep and 30,326,000 pigs (Table 2.3). France and West Germany together contained half of all cattle and pigs, although individual herds were usually very small, and the UK contained a similar proportion of sheep. Not surprisingly, more than half of all meat and milk produced in Western Europe

Table 2.3 Livestock numbers, 1950, and changes, 1950–80 (per cent)

	Cattle		Pigs		Sheep	
	1950* ('000)	1950–80 (%)	1950* ('000)	1950–80 (%)	1950* ('000)	1950–80 (%)
Belgium	2,017	+ 43	1,142	+ 336	165	− 45
Denmark	2,998	− 2	2,829	+ 238	61	− 11
France	15,606	+ 51	6,582	+ 60	7,499	+ 72
W. Germany	10,927	+ 37	9,563	+ 133	2,046	− 42
Greece	763	+ 22	549	+ 72	6,978	+ 80
R. of Ireland	4,211	+ 46	615	+ 81	2,419	− 2
Italy	8,281	+ 6	4,026	+ 119	10,187	− 1
Luxembourg	133	+ 63	96	− 17	4	+ 25
Netherlands	2,659	+ 89	1,561	+ 543	404	+ 121
UK	10,277	+ 30	3,363	+ 131	19,945	+ 9
EC	57,872	+ 36	30,326	+ 151	49,708	+ 24

Note: *Average for 1947–52.

derived from France and West Germany. Once again the superiority of north over south was clearly apparent.

Transformations

Tradition dominated over innovation and muscle power over mechanization at mid-century but important transformations were already under way in a handful of areas in Western Europe and these heralded the kinds of change that were to affect the rural environment in the future. Four examples, from the UK, Italy, Greece and the Netherlands, will be examined here.

During the 1920s and 1930s the UK had experienced an intense phase of urbanization, with agricultural depression, low land prices and rudimentary planning controls facilitating the conversion of 360,000 ha from rural to urban (or more precisely suburban) uses between 1925 and 1939 (Wibberley 1961–2). Indeed, such a rapid rate of land-use change has not been experienced since then in the UK. Conditions changed dramatically during the Second World War when commodity imports were severely curtailed and an impressive drive was initiated to revitalize domestic agriculture and greatly increase the volume of food produced at home. A remarkable expansion in the area under allotment gardens had no small part to play (Thorpe 1975).

The early days of war saw the start of a campaign for ploughing up permanent grassland to expand arable cultivation, together with strong financial incentives for intensifying output (Tracy 1982a). In this atmosphere of siege, Britain's ploughland increased from 5,200,000 ha in 1939 to 7,200,000 ha in 1944, while permanent grass fell from 6,800,000 ha to 4,400,000 ha. Crop yields rose dramatically, in response to improved cultivation techniques, better seed and artificial fertilizers, and self-sufficiency in food supply was boosted in a way that was inconceivable under free market conditions. Of course, the plough-up campaign was temporary and exceptional and ran parallel with food rationing and a marked reorientation of diets away from animal products, but the experiment showed what heavily subsidized scientific agriculture could achieve and hinted at some of the environmental changes that might be expected in later decades as agricultural intensification occurred in many other parts of the Ten.

On the mainland of Europe important programmes of marshland drainage and land reclamation were adding new technology to long-established methods to transform a number of rural areas. In the forty years following legislation in 1882, 1,000,000 ha were converted to productive agriculture and forestry in Northern Italy in accordance with the concept of *bonifica*, which embraced drainage, irrigation, deep ploughing, erosion control or reafforestation according to local circumstances (Mioni 1978). Between 1922 and 1940 a further 1,500,000 ha were improved but this fell short of Mussolini's grand design for 8,000,000 ha, since Fascist *bonifica* formed part of a massive job-creation programme in depressed times and had clear political motivation (Gay and Wagret 1979). Its prime example involved the Pontine Marshes, an area of 80,000 ha covered with marsh, scrub and rough woodland that was flooded every winter. Malaria was endemic among the 3000 inhabitants, and animals that grazed in the marshes were weakened by disease. After a detailed survey, reclamation started in 1931, involving clearance of rough vegetation, cutting drains, deep ploughing to break up hardpan, and the eradication of malaria.

By 1936 the work was complete, with 500 km of new roads, 2000 km of canals, and 48,000 ha deep ploughed, fertilized and ready for cultivation. Some 2500 farmhouses had been built and by 1938 the population had risen to 60,000. Holdings varied in size from 10-12 ha in fertile areas to 15-25 ha on poor sandy soils. Aqueducts brought irrigation water and windbreaks surrounded the new fields clothed with crops of wheat, sugar beet, fodder and cotton. The wetland environment had gone and had been replaced by a new cultural landscape that was created through a combination of intensive labour and modern methods in less than a decade. Large sections of the Roman Campania were also transformed and lowland schemes were started in Sardinia; however, the rural problems of the hills and mountains and of the Mezzogiorno in general were not really tackled prior to the 1950s (Houston 1964). Above all else, the Pontine Marshes provided a spectacular advertisement for the Fascist regime and formed an important precedent for things to come (Gentilcore 1970).

On quite a different scale, a modest start had been made on irrigation and marshland reclamation in various parts of Greece during the interwar years, with government grants being allocated in the late 1930s. However, much damage occurred during the German occupation in wartime and work had to be started anew in 1945, and more specifically in 1947 when US economic aid to Greece began. In the following ten years substantial improvements were carried out to enhance the country's desperately poor agricultural resource base. Some 389,000 ha were protected against floods, 279,000 ha of marsh or low-lying soil drained, 48,000 ha irrigated and a further 73,600 ha prepared for it, thus helping the country's total irrigated surface to reach 380,000 ha in 1959. In addition, 80,000 ha of scrubland were cleared for cultivation and large areas of alkaline and saline soils improved by deep ploughing and application of artificial fertilizers. The start of modern agriculture activity in Greece effectively dated from this time (Medici 1969).

Not surprisingly the most imaginative and integrated scheme for rural transformation was not to be found in Mediterranean Europe.

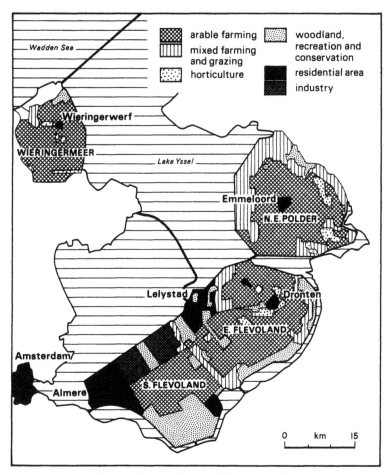

Figure 2.2 Land use in the Ijsselmeer polders (after Meijer 1975)

After centuries of polder reclamation the Dutch had started on the formidable task of reclaiming the Zuyder Zee in 1927, draining first a 40 ha experimental polder and the Wieringermeer polder (20,000 ha) and then working for five years on the North Sea barrier dam (Meijer 1975). By 1942 the north-east polder (48,000 ha) had been pumped dry but attention soon returned to Wieringermeer which was flooded by German troops (Figure 2.2). The East Flevoland polder (54,000 ha) was reclaimed between 1950 and 1957 and in 1959 drainage engineers turned to the South Flevoland polder (42,000 ha) which was reclaimed by 1968. Old-established drainage techniques were successfully brought up to date with the help of important financial backing from the Dutch state. The areas to be drained were ring-diked and then pumped dry. As soon as the land had fallen reasonably dry, reeds were sown to accelerate the drying-out process through evapotranspiration. Once land had become relatively firm the reeds were cut or burned and cultivation began on the new farmland, ribbed initially by open ditches that were later replaced by field drains of earthenware or perforated plastic. Roads, bridges, farms and settlements were installed by the state authorities which supervised farming during the first few years. Land was duly leased to farmers chosen because of their competence or their need to replace properties lost through urban expansion or road construction.

The new polders enabled the land surface of the densely populated Netherlands to be extended by 164,000 ha (an increase of 5 per cent) and a set of completely new rural landscapes and settlement patterns to be created in just a few decades (Van Lier and Steiner 1982). In addition, they provide a land specimen of much more general problems of land allocation and service provision that are encountered in long-settled areas in many parts of Western Europe. Wieringermeer polder was divided into a regular grid of elongated 20 ha parcels that could be assembled into different-sized holdings. Farmsteads and barns were built along a network of rural roads and four small villages were constructed to provide shops and other services. Tree planting along roads and around settlements provided shelter and an element of diversity in an otherwise strongly horizontal landscape. The villages proved to be too small and too numerous to provide more than very basic services and their hinterlands overlapped, but because of the sectarianism of Dutch society each was provided with two or three churches and schools.

Planners sought to avoid such problems in the north-east polder and, following the inspiration of Christaller's (1933) research on central places in Southern Germany, they designated one large settlement (Emmeloord) at the intersection of two road axes and five smaller villages that were served by a roughly circular road system. Farmhouses were constructed adjacent to roads in the midst of fields but accommodation for most farmworkers was built in the villages. A survey undertaken elsewhere in the Dutch countryside suggested that 5 km was the maximum desirable distance between farm and village in the car-less days of the 1940s, hence the number of villages on the north-east polder was increased to ten, with each being planned to house 1000–2000 persons. Use of prefabricated materials gave farmhouses a remarkably similar appearance throughout the polder, although shelterbelts and small plantations of woodland provided some diversity in the landscape. Emmeloord now houses over 15,000 people but only three villages have more than 1000 inhabitants and the remaining seven house only 500 people apiece. Mechanization and farm enlargement have trimmed back the number of workers needed in agriculture and widespread car ownership has allowed much greater distances to be travelled to obtain goods and services.

Initial plans for East Flevoland placed strong emphasis on providing agricultural land and included a similar settlement pattern to that on the north-east polder (namely ten small villages and one sub-regional centre as well as Lelystad to serve the whole Zuyder Zee region), although farm sizes were to be larger. All these ideas were soon revised; the number of villages was reduced to two, and less land was

Table 2.4   Land use in the Dutch polders

		Land use (%)			
	Area (km²)	*farm-land*	*woods, conser-vation*	*housing*	*roads, water-courses*	*Average farm size (ha)*
Wieringermeer	193	87	3	1	9	36
North-east polder	469	87	5	1	7	26
East Flevoland	528	75	11	8	6	40
South Flevoland	430	50	25	18	7	n.a.
Netherlands	33,812	68	13	19		12

Source: H. Meijer (1975).

devoted to farming, with much more emphasis being placed on woodland and space for recreation and housing (Table 2.4). South Flevoland displays an even lower percentage of its surface devoted to agriculture and has been planned to meet a wide range of objectives of which farming is simply one. These matters are not unique to newly reclaimed land in the Netherlands, rather they have been encountered in one form or another throughout the Ten during the past three decades as farming has been modernized, established rural population has been redistributed, and new pressures for residence, recreation and conservation have been exerted in the countryside.

3 Depopulation

Processes of change

A traditional role of the countryside has been that of a vast 'nursery of mankind' where human pressures on local food supplies were kept in check by various forms of outmigration. Countryfolk in their millions left farms and villages to provide much of the human potential that was vital for city populations to reproduce and grow, while at the same time supplying labour that was necessary to build urban environments and to operate urban factories. Normally these rural-urban movements were permanent, although there were also many historic examples of rural dwellers migrating to town on a temporary or seasonal basis. This centuries-old model of relationships between cities and their rural hinterlands retains its validity in the second half of the twentieth century since definitive rural-urban migration has continued in recent decades and has in fact been encouraged by planning policies which incorporate growth-pole principles for concentrating investment and employment at a limited number of urban locations. But this model is now just one among several, since the broad process of urbanization has taken on new characteristics which have profound implications for patterns of spatial interaction, for the social structure of settlements in the countryside and for the use of land (Hochas 1977).

Despite the many processes that contributed to the repopulation of certain rural areas, many stretches of countryside in the Community continued to experience a loss of population in the decades after 1950 (Moore 1979). Indeed the finer the scale of scrutiny, the more clearly rural depopulation can be identified since it has come to be associated with settlements at the lower end of the size hierarchy that are located in areas beyond 'regional cities', are away from main transport axes, and are not particularly attractive for commuters, retired migrants

or other replacement populations (Kayser 1969, Leardi 1973). By
contrast, larger rural settlements and small market towns have often
managed to function as 'anchoring points' or 'key settlements' and
have retained their population to a greater degree (Béteille 1977).
Depopulation is certainly not just confined to marginal rural areas but
is also encountered in prosperous farming districts where mechanization and other farm-based changes have generated redundancy but
sufficient alternative off-farm employment has not been available
(Drudy 1978). However, it is fair to state that upland areas have been
particularly prone to population loss (Schéma Général d'Aménagement
de la France 1981). For example, mountainous areas of Italy lost an
average of 30 per cent of their inhabitants between 1950 and 1980 and
village population declined by half in some stretches of the Apennines
(Bethemont and Pelletier 1983). In some parts of Western Europe
losses from villages and surrounding rural areas were so consistent that

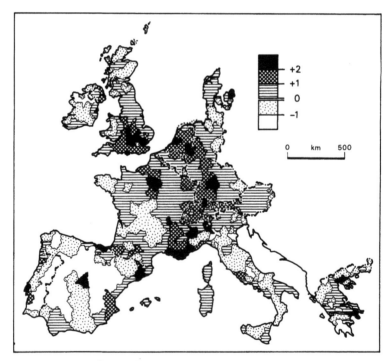

Figre 3.1 Population change in Western Europe, 1951–71 (percentage
change each year)

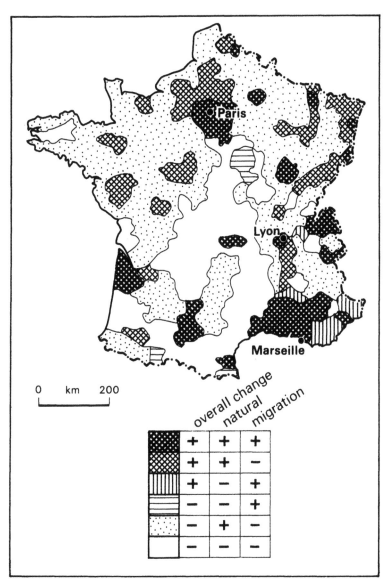

Figure 3.2 Population change in rural France, 1962–8 (after Fel 1974)

they affected provincial trends during the 1950s and 1960s, forming a great arc of population decline that extended through Ireland, parts of Scotland and Central Wales, to sections of Central France and much of Italy and Greece (Figure 3.1). In fact, it continued beyond the limits of the Ten into Spain and Portugal which contained districts with the highest rates of annual loss anywhere in Western Europe. In contrast, countries such as West Germany and Belgium, endowed with widely dispersed networks of dynamic urban centres and precocious systems of daily commuting, managed to retain people in the countryside to a much greater degree.

Absolute reductions in numbers of rural residents are produced by several interrelated processes. The number of deaths in a rural area may exceed births during a specified period and 'biological depopulation' will occur if net inmigration is insufficient to compensate such a trend. This kind of situation normally occurs only after protracted out-migration of young people has distorted the age/sex structure of the local population and rendered it unable to reproduce itself. Such a condition did, however, characterize extensive areas of the Massif Central and south-western France during the 1960s (Fel 1974) (Figure 3.2). Many settlements in these regions had achieved their peak populations one hundred or more years earlier!

Net outmigration is directly responsible for conditions that have been distinguished as 'non-occupational' and 'occupational de-population' (Pinchemel 1957). The first type affects areas where employment opportunities and/or social conditions are particularly depressed and large numbers of residents, regardless of their specific occupation, choose to move away. 'Occupational depopulation' affects members of clearly defined employment categories, such as rural craftsmen, landless labourers or tenant farmers. In both instances net outmigration serves to reduce the significance of any natural increase that may occur and with the passage of years thereby produces an overall loss in population. For example, powerful outmigration cancelled out natural increase across large areas of north-western, Northern and Eastern France during the 1960s (Figure 3.2).

Rural exodus

The reasons for rural residents choosing to leave the countryside are complex in the extreme and may be explored at varying scales, ranging from considerations of broad economic trends, through socio-economic conditions in particular settlements, to the detailed perceptions of places and opportunities held by individuals. Starting at a high level of generalization, it is clear that different sectors of the ten economies have experienced varying degrees of dynamism during the years since mid-century. Post-war reconstruction of homes, factories and transport services, building new housing to make up for time lost in the interwar years, and national brands of 'economic miracle' all stimulated powerful demands for labour in the building, public works, manufacturing and tertiary sectors from 1950 until the early 1970s. A proportion of these demands was met by workers released by (or simply not absorbed in) traditional rural activities, although much labour migration was also required from sources beyond the EC. Long-established crafts and services which had flourished in the countryside to satisfy local needs were progressively outbid by urban manufacturers and suppliers. Certainly the number of rural craftsmen working wood, metals, leather, fabrics and other materials had already declined markedly but a residual group still remained at mid-century. Very many of them were vulnerable to external forces and were soon to lose their livelihood just as their predecessors had done in earlier decades. Rising rates of car ownership enabled many members of the remaining rural population to drive to their nearest market town to obtain a range of goods and services, thereby reducing the viability of village shops and mobile retailers, and this tendency ran parallel with a trend for administrative reorganization and the closure of primary schools and post offices in small settlements in most countries of the Community.

In addition, there has been a remarkable contraction in employment on farms which has long been associated with low rates of pay, relatively hard working conditions and often poor housing as well (Laurent 1975). In 1982 only 8,164,000 people were engaged in agricultural work in the EC, representing no more than one-third of the total recorded thirty years earlier (Kouloussi 1983). In other words there were 15,631,000 fewer farmers and agricultural labourers – a figure somewhat greater than the total population of the Netherlands in 1981 (Table 3.1). A sharp reduction in the number of young men in farmwork contributed to a general ageing of the remaining agricultural population and to a rather greater emphasis on its female component in some countries (Barberis 1968). Below-average rates of loss in agricultural employment involved advanced countries, such as Denmark, the Netherlands and the UK, which already had fairly low

Table 3.1   Employment in agriculture, 1950 and 1982

	Total numbers ('000)		Decline	% of workforce	
	1950	1982	(%)	1950	1982
Belgium	368	106	− 71.2	11.3	2.9
Denmark	486	206	− 57.6	23.0	8.5
France	5,438	1,758	− 67.7	28.3	8.4
W. Germany	5,020	1,382	− 72.5	24.7	5.5
Greece	1,962	1,083	− 44.8	52.0	30.7
R. of Ireland	580	196	− 66.2	47.0	17.3
Italy	8,261	2,545	− 69.2	41.0	12.4
Luxembourg	32	8	− 75.0	24.0	5.1
Netherlands	533	248	− 53.6	14.1	5.0
UK	1,115	632	− 43.3	6.0	2.7
EC	23,795	8,164	− 65.7	n.a.	7.7

proportions of their workforce engaged in farming at mid-century, but also Greece which had just over half of its workforce in its agricultural sector in 1950 and retained 30 per cent there three decades later. The Republic of Ireland (with 17 per cent of its workers in farming) and Italy (12 per cent) also kept large numbers on the land in 1982, thereby maintaining the Community's agricultural periphery in which alternative forms of full-time employment were not readily available (Figure 3.3). However in the last ten years the rate of annual loss in farm work has declined in most parts of the EC because of sharply reduced job opportunities in other sectors of the economy.

Just over half of the agricultural workforce in 1981 was made up of farmers themselves, with the proportion rising to a remarkable 82 per cent in Denmark (Table 3.2). Family members represented one-third of the total and that figure was greatly surpassed in West Germany, Italy, the Republic of Ireland and doubtless also in Greece, although comparable statistics are not available for that country. The UK was exceptional in having over one-third of its agricultural workforce made up of non-family workers, more than double the average of the Community as a whole. Nevertheless, the number of full-time workers in England and Wales has shared in the general trend, falling from 570,000 in 1950 to 163,000 three decades later and with current losses standing at 7000 each year (Wallace 1981).

The 15,631,000 reduction in numbers involved in farmwork

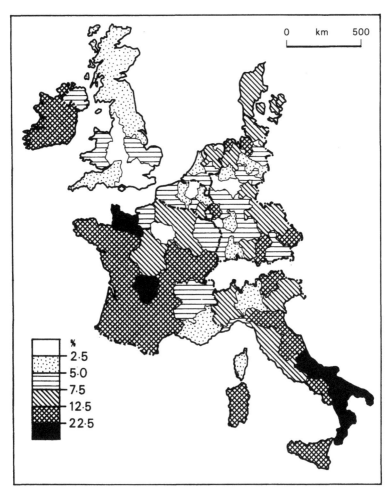

Figure 3.3 Proportion of the workforce in agriculture, 1980 (per cent)

represented the overall outcome of various processes of change but only some of these contributed to rural depopulation. There were 4,000,000 fewer farmers in 1982 than three decades earlier, as operators chose to quit working the land, retired or died with no successor to continue operating their property. This could be sold or leased to a new occupant, might be used to enlarge surrounding holdings, or might simply be left uncultivated. But the greater part of the reduction involved family members and non-family workers who

Table 3.2   Composition of the agricultural workforce, 1981 (per cent)

	Farm occupiers	Family members	Non-family workers
Belgium	75	20	5
Denmark	82	4	14
France	52	36	12
W. Germany	52	39	9
R. of Ireland	50	42	8
Italy	49	42	9
Luxembourg	45	55	0
Netherlands	56	29	15
UK	54	10	36
The Nine	52	35	13

chose to seek off-farm employment and whose presence on the farm was ·less critical as many aspects of farmwork were gradually transformed through mechanization. In some instances they migrated to urban areas in their home region or home country in search of suitable alternative work if this was not available in their own settlement or within easy access of it for journey-to-work purposes. In addition, large numbers of former farmworkers travelled from Greece and the Mezzogiorno to city jobs in other parts of the EC. Other people who left farmwork managed to find alternative employment in their home area and continued to reside in their village. In such instances the agricultural exodus was far from being identical with the rural exodus.

Beneath this range of broad economic factors there were marked variations in material and social conditions in specific settlements which influenced migration decisions. Important progress has been made in improving facilities in rural areas throughout the EC in the past three decades, but inferior housing, absence of piped water supplies, electricity and mains drainage, and lack of adequate public transport and local facilities for shopping and education undoubtedly influenced many rural dwellers to leave the countryside. However, there was no simple correlation between paucity of rural facilities and intensity of outmigration. Rather more elusive issues, such as variations in community spirit, the quality of leadership and the general conviviality of social conditions in particular settlements, also played their part. For example, strong patriarchal control of village society proved very stressful for many young unmarried Greeks who opted to leave their home area in search of opportunities in the city (Carter 1968).

The sex, age, education and precise residence of potential migrants were other important social influences. The chance of inheriting property and running a farm may have provided sufficient reason for many young men to remain on the land but farm girls were less likely to inherit and may have been strongly influenced to move away by the harsh working conditions that their mothers had to endure. In any case most rural areas failed to offer adequate opportunities for off-farm work for females and many young women saw their future in an urban area rather than on a farm. They tended to migrate at rather younger ages than their male counterparts who, if they chose to leave, often postponed migration until after military service or until they had decided to search seriously for a wife. Traditions concerning the eligibility of people for marriage also played an important role in stimulating migration away from home communities, notably in rural Greece where strict consanguinity constraints proscribed marriage between relatives and thereby encouraged young people to search further afield for a partner (Burgel 1972). The decision to move was often associated with such critical phases relatively early in the life cycle, including leaving school, searching for a first job, settling down after military service, looking for a partner, or finding suitable accommodation and employment to support a young family.

Variations in schooling were critically important, with the higher the level of education the greater the likelihood of young people moving to town in order to find work that allowed them to make use of their training (Mathieu 1974). Many rural parents strongly encouraged their children to pursue their studies so that they might train as nurses, teachers or white-collar workers and have marketable skills that were always believed to be of value for urban employment. In some instances parental encouragement to continue with education was directed more strongly towards girls than boys, since young men might have a chance to work on the farm (Hannan 1969, 1970). Traditions regarding land inheritance and duty to one's parents also influenced decisions to migrate; some parents advised their children to move and improve their lot elsewhere, whereas others encouraged at least some of them to stay in order to work the land.

Ultimately there were intensely personal and psychological matters that differed for each potential migrant. Variations in satisfaction and ambitions for the future regarding work and general lifestyle were partially a response to an individual's character and intelligence but were also linked to the range and quantity of information that had been available for sampling. Formal education, radio, television, newspapers, contacts with friends and family living elsewhere, and holiday travel provided images of other environments with perceived advantages that might outweigh life in the countryside. Out-migration, for whatever set of reasons, inevitably produced serious imbalances in the settlements left behind.

Implications

As well as contributing to new tensions in urban destinations (which are beyond the scope of the present discussion) rural/urban migration has had profound implications for rural departure areas. These are affected with respect to size and structure of their resident population, pressure on land resources for farming and other uses, and provision of local services for those who remain. Deprived of their young people, residual populations become increasingly elderly, and needs for peripatetic medical and social services intensify but can rarely be satisfied in thinly populated areas where distances between clients are great and the costs of providing such services are exceptionally high. The cumulative effects of decades of outmigration and associated biological depopulation in some localities has produced a vast, almost unbroken stretch of 'elderly' territory extending from the upper Seine valley to the Pyrenees and from Poitiers to Nîmes where rural *cantons* had more than 20 per cent of their population aged 65 years or over in 1975. In 1954 the population of only 126 rural *cantons* throughout France had passed that threshold; by 1962 the total was 173 and continued to rise to 400 in 1968 and 714 in 1975, when 19 per cent of all residents in the French countryside were aged 65 or over and 7 per cent were over 75 (Paillat and Parent 1980). Similar features, albeit on a smaller scale, already characterize many other rural sections of the Community and are being intensified by the growing trend for retirement migration to the countryside and the fact that West Europeans are living longer.

A second major implication of rural depopulation is the reduction of commercial, administrative and other services in many small settlements and thinly peopled areas. This trend derives in part from the shrinkage of demand which depopulation carries with it but also stems from the changed lifestyles of many rural residents and from official policies for 'rationalizing' the provision of educational, medical and other services at a limited number of 'key settlements' (Cloke 1979, 1983). Fewer potential customers means that business turnover will be less and this basic truth has led to the closure of village stores, sub post offices, pharmacies, public houses, cafés, and garages on minor roads throughout rural Europe. Individual enterprises with limited financial turnover are often left without successors once the existing owner or shopkeeper has retired or died, and small unremunerative outlets which form part of larger retail, brewing or garage chains are particularly liable to closure. Village surgeries are disappearing rapidly, as the trend develops for group practices in market towns, and small rural hospitals are being closed in favour of larger, more centralized facilities that serve wide hinterlands. Many of the rural residents who remain have cars of their own and can travel to the nearest market town for shopping or to the main road to obtain petrol. As rates of car ownership continue to rise this fact becomes all the more salient. Currently 66 per cent of journeys in British towns are by car but the figure reaches 81 per cent for residents of the countryside. Difficult terrain and poor driving conditions may provide a hint of reprieve for some villages and small towns in mountainous regions but there is clear evidence that patronage of lower-order service centres is declining rapidly in most parts of rural Europe (Barbier 1972). This tendency to drive elsewhere for goods and services obviously reduces the viability of outlets that cling on tenaciously and operate to low profit margins and ultimately contributes to a vicious downward spiral of local service provision (Van der Haegen 1982).

The same point applies to the contraction and in some cases total disappearance of any kind of public transport in many rural localities in the northern parts of the EC. From their maximum extent during the late 1950s, services have been cut back massively, although village bus services seem to operate to a greater degree in Mediterranean Europe where rates of car ownership are rather lower. Acquisition of deep freezes by a growing number of more affluent rural dwellers also reduces the need for ready access to retail outlets. However a residual need for shops, services and public transport is still to be found among clearly recognized groups living in the countryside, notably the poor, the sick and infirm, and the elderly. In many instances they are the most acute victims of rural deprivation.

The closure of village schools is another response to the general decline of population and progressive ageing of remaining residents in depopulated parts of rural Europe. This has been accentuated by regulations in many countries that specify pupil thresholds for the maintenance of country schools. For example, legislation in the UK in 1944 abolished basic village schools for all children aged between 5 and 13 years and distinguished between primary facilities, which would be available locally but not in every village, and more centralized secondary schools, which would operate at a smaller number of places and to which children would travel individually or be taken by bus each day. Many village schools were closed subsequently in localities where numbers of primary pupils were small. Reports in 1967 commented disparagingly on the educational quality of small schools, recommended a minimum size of three teachers and at least fifty pupils (unless there were exceptional local circumstances), and advocated more closures. These duly followed, as birth rates in depopulating areas continued to decline and local authorities attempted to engineer financial savings.

Very similar theory and practice were being evoked in other countries, including the French-speaking southern part of Belgium where legislation in 1975 defined pupil thresholds in relation to population densities and had the effect of reducing the number of village schools from 4200 in 1972 to 2375 in 1979 (Christophe 1982). However, at the very same time rural planners in France were rejecting this kind of policy and were advocating a much more cautious approach (see Chapter 8). They recognized that such closures had profound implications since the village school, together with the church, often acted as a kind of focus for local people in thinly populated districts, whether they had children attending there or not. Once the school had been closed that critical point of identity was gone. The few families with young children who remained might decide to move to larger villages or towns where schools were available, and there would certainly be little attraction for young families to move into school-less villages and hamlets. The fact could have serious negative consequences for any attempt to promote additional housing or new forms of employment, such as small-scale light industry, in villages that did not have primary educational facilities (McLaughlin 1976).

All these types of closure interact upon each other to generate a cumulative effect, and give rise to additional withdrawals of service.

For example, a survey of twenty-six rural parishes in Scotland noted that not only had the number of shops and post offices fallen by half since 1960, but the number of doctors and primary schools had declined by over one-third (Scottish Consumer Council 1982). The same survey highlighted the fact that people living in such areas tend to have low wages and yet have to pay dear for the goods and services that they need. Average wages in those twenty-six parishes were 10 per cent lower than the average for Scotland (which was already 5 per cent lower than the GB average), while prices for everything were higher: 14 per cent more for food, 9 per cent more for petrol, and 12 per cent more for other goods. Studies of retail prices for other parts of the UK came up with comparable results, showing that the price of 'the standard shopping basket' of goods was between 2 and 14 per cent higher in village shops than in town supermarkets. Features such as these make depopulating rural areas surprisingly expensive locations in which to live and remarkably difficult places for those who cannot afford a car or for some other reason do not have transport of their own. In reality, there is not much financial advantage for a rural household to shop in town since slightly lower retail prices are offset by the extra cost of public transport or running a car and meeting parking charges. The existence of such problems in thinly populated areas is gradually being appreciated and more flexible approaches to service provision are being advocated and are actually being implemented in some areas of the EC.

A third implication of depopulation is to reduce pressure on housing, land and other local resources that may permit property to be sold and surrounding farms to be enlarged. But in the poorest and most remote rural areas of the EC there may be no takers and 'villages and isolated rural houses become rather like collections of snails with the shells left behind' once the inhabitants have gone (Wibberley 1975, 92). Formerly cultivated land is simply abandoned, being owned by urban relatives who are long distanced from local farming stock and have not the slightest interest in working their land. In the higher parts of the Mezzogiorno and Central Italy, sections of the Alps and remote areas of the Massif Central, more or less deserted hamlets and farmsteads are now surrounded by crumbling field walls, and plots which once produced arable crops or decent yields of fodder have reverted to scrub or rough woodland. In mountain locations such as these the 'frontier' of settlement and agricultural occupation has simply retreated downslope (Lichtenberger 1978). King and Young

(1979) have written evocatively of 'the death of a human landscape' on the depopulating Aeolian islands to the north of Sicily, where once fastidiously tended terraces are disappearing under the invading natural vegetation. Other evidence of past activity is progressively being obliterated by the creeping *maquis*; the houses are locked and stand derelict since the migrants never come back. Similar experience is widespread throughout upland Europe.

Land abandonment has been particularly severe in the hills and mountains of Italy in response to the powerful rural exodus that operated between mid-century and the early 1970s. The agricultural surface declined by one-eighth nationwide between 1961 and 1981 and arable land by one-quarter, with retreat being particularly severe in the Apennines (Ruggieri 1976). Numerous hazards have arisen here and in other uplands once cultivation has ceased on sloping land which soon becomes covered with rough grass or low shrubs. Abandoned terraces and other slopes suffer serious damage from landslips, mud torrents and accelerated runoff, as well as acting as avalanche paths in winter and perhaps being devastated by uncontrollable fires in summer. Encroachment of eroded soil on to neighbouring or low-lying plots of land creates other complicated problems in the realm of legal liability. In 1952 and again in 1971 special legislation was introduced to encourage local authorities in Italian mountain areas to devise land-use plans and practical ways of conserving soil resources in depopulated districts (Bagnaresi 1978). The responsibility for making proposals for settlement planning, reafforestation of unstable or infertile soils, and reclamation of more fertile areas now lies with those local communities but intricate patterns of landownership and lack of interest among residents and absentee owners has restricted their practical achievements (Abrami 1978).

Comparable environmental degradation has resulted in other abandoned mountain districts in France, West Germany and Greece. Areas in the Peloponnesus and on some of the islands reached their peak population before 1900 and virtually the whole Greek countryside has experienced demographic decline since then (Figure 3.4). During the 1940s population loss in upland Greece and frontier districts was due to war deaths and migration of mountain people to more secure lowland areas. Migration downslope and to overseas destinations has continued powerfully in recent years, with rates of loss being particularly strong in thinly populated mountains and on small islands that the tourists have not discovered, where houses are in ruins, villages are

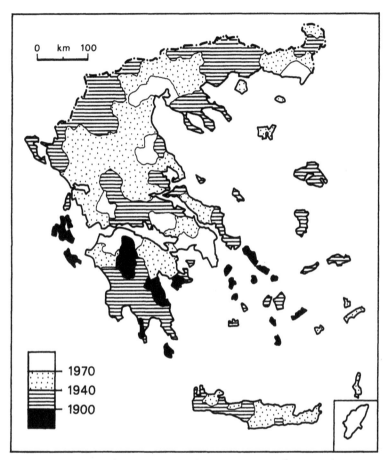

0 km 100

1970
1940
1900

Figure 3.4 Dates of maximum population in Greece (after Burgel 1981)

abandoned, terrace walls have collapsed and farmland lies idle. Abandoned patches of land are also found in the midst of cultivated fields in lowland areas because many migrants retain ownership of their plots as a kind of insurance policy against hard times, even though their land is not used on a regular basis. Sometimes it may be worked casually by relatives or grazed by livestock from neighbouring properties but most abandoned plots are simply left to harbour weeds and vermin which have a damaging effect on surrounding plots. Absentee ownership also hinders plot amalgamation and the creation of viable farm units (Wagstaff 1968). Exactly the same kinds of ecological and structural problems are encountered in rural areas elsewhere in the EC which have considerable stretches of abandoned farmland. Depopulation still continues in some sections of countryside in the Ten in the 1980s but with reduced intensity following the collapse of alternative job opportunities in urban centres since the mid-1970s. In addition, a range of reverse migration flows are now in operation and are working to repopulate extensive stretches of countryside in a variety of controversial ways.

4 Repopulation

Commuting

Since mid-century a number of processes have operated more forcibly than before in order to change the functional relationship between town and country and thereby allow certain types of rural area to be repopulated (Lewis and Maund 1976). The countryside has become fashionable among many city dwellers who have responded to the attractions of nature, open space and the wholesomeness of rural life (Gilbert 1978). In addition, the widespread tendency to decentralize certain types of employment from major cities to smaller towns has generated important repercussions in surrounding stretches of countryside. Daily commuting, retirement migration, visiting the country for recreation, and retreating from an urban way of life to a smallholding each had important precedents in the first half of the twentieth century but have become particularly significant since 1950, spurred on by the fifteen-fold rise in car ownership in the EC over the past three decades which has provided more West Europeans with a greater degree of individual spatial freedom than ever before. To express the magnitude of the transport revolution in another way, West Germany with 23,731,000 cars now contains four times as many cars as all ten countries together in 1950 (Table 4.1).

There are two quite distinctive aspects to the daily commuting phenomenon. One involves rural residents, often from farming backgrounds, who take on work in a city or town and, rather than uproot themselves and their families to urban accommodation, choose to live in their home village or farmhouse and travel to work each day. They may even contrive to farm their land on a part-time basis (see Chapter 5). As far back as 1869 a policy started for providing Belgian manual workers with railway season-tickets at one-sixth of the regular price and this practice was enhanced by a second type of cheap

Table 4.1   Cars in the European Community, 1950 and 1981

	Number of cars ('000)		Cars/thousand population	
	1950	1981	1950	1981
Belgium	274	3,206	32	325
Denmark	118	1,367	27	267
France	1,600	18,800	38	349
W. Germany	598	23,731	13	385
Greece	20	911	5	94
R. of Ireland	40	778	10	225
Italy	342	18,603	7	325
Luxembourg	10	133	35	365
Netherlands	139	4,609	14	323
UK	2,317	15,910	47	283
EC	5,458	88,049	25	324

workingman's ticket during the 1930s (Kormoss 1976). By 1950 daily commuting had become a thoroughly ingrained tradition, endowing Belgium with the highest degree of separation between home and workplace in any part of Western Europe (Dickinson 1957). Daily migrants travelled to Brussels from almost every part of the country and other major cities also commanded sizeable commuting hinterlands. Currently, half the working population in Flanders is made up of commuters and that region has undergone a remarkable degree of dispersed urbanization (Van der Haegen 1982).

Conditions were rather unusual in West Germany where rural/urban commuting rose dramatically in importance in the years immediately after the Second World War (Dickinson 1959). The increase was boosted by the forced influx into villages of a great number of refugees from eastern parts of Germany as well as evacuees from war-torn cities whose labour was required in urban centres each day as the German 'economic miracle' started to get under way after 1948 (Klöpper 1971). Acquisition of motor-assisted bicycles during the 1950s contributed to a widening pattern of commuting in the Netherlands and the same role was played by motor scooters in Northern Italy (Turton 1970). Works' buses operated special services around mines and industrial plant in many parts of the Ten. For example, the Shannon industrial estate in Western Ireland enrolled a large commuting labour force, some of whom travelled up to 50 km each way in works' buses every day (Küpper 1969). Cheap oil prices and the massive increase in car ownership during the 1960s and early 1970s strengthened the pattern of rural-urban journeys to work and endowed commuters with greater ease of travel than ever before. Major new industrial schemes, such as the Taranto steelworks in Southern Italy, soon established functional hinterlands in the surrounding countryside as large numbers of employees travelled to work from their agricultural holdings each day (Lieutaud 1982). In addition, the conditions of the 1960s and early 1970s favoured a second kind of commuting to develop.

This has involved urban dwellers in their millions moving to homes in small towns and villages beyond the continuously built-up fringe of suburbia, which in some countries has been stabilized by official 'green belt' policies (Johnson 1976, Munton 1983). As Pahl (1965) insisted two decades ago, the outer part of a metropolitan region represents a frontier of social change which sweeps across settlements and sections of countryside, modifying their social structure, physical appearance and functional relationship with major cities. Four critical components were identified in this broad process, namely inmigration of large numbers of former city-dwellers, commuting to urban employment by a sizeable proportion of residents, segregation of housing areas in the affected settlements, and a collapse of geographical and social hierarchies. Incomers to the countrysides of Western Europe are far from uniform in social characteristics, ranging from well-to-do professional workers to less affluent people who may be thought of as reluctant commuters (Cohou 1977). Motives for moving into the countryside also vary greatly, with some incoming families being able to buy large houses in their own grounds that afford them the living space they desire and enable them to raise children in a pleasant 'village community' or 'small town atmosphere' (Connell 1978). Other incomers have much more limited finances and are attracted by the availability of cheaper types of housing which are more pleasant than comparably priced accommodation in the city. Of course, they also have to take into account the cost of lengthy journeys to work each day.

All types of incoming family are likely to live substantially different lifestyles from the greater share of old-established villagers and farmers, whom they may eventually outnumber. For example, notable differences in household equipment, degree of female employment, and general levels of income and education have been recognized in Southern Belgium, not only between settlements dominated respectively by commuters or farmers but also between commuter villages which house industrial employees and those that cater for white-collar workers (Mougenot 1982). Commuting by train, private car or motorcycle is a necessary daily ritual for most members of the incoming workforce since precious few jobs are available in these invaded settlements. Spatial segregation occurs on a local scale as different types of incomer occupy existing housing of varying size and quality set in more or less attractive locations, and as estates of variously priced new housing are constructed around old village cores. Long-held distinctions between small towns, villages and hamlets lose much of their significance as the tide of incomers converts them all into functional parts of the 'dispersed city' (Mathieu 1972). In many instances any semblance of a cohesive rural 'community', with its layered components from agricultural labourers through craftsmen to rich landowners, is lost and conditions become polarized between combinations of established and incoming households which pursue either middle- or working-class lifestyles. For example, incomers and large farmers living in commuter villages around Rouen in Northern France displayed quite different patterns of mobility, behaviour and urban orientation than did small farmers and agricultural labourers who continued to live in much more local worlds (Guermond 1973).

None the less, commuter villages contain significant subgroups of people who experience quite different needs and problems. The concerns of retired folk for local shops and medical services are far removed from the worries of young families about the provision of local schooling. Residents who possess no car of their own and have to rely on buses or trains view issues relating to public transport in a very different way from highly mobile households with two or three cars apiece. Rich incomers tend to favour conservation of the fabric and appearance of the old village, whereas less affluent residents may welcome more construction and development if that offers the prospect of additional jobs and services in the settlement. After decades of stagnation, many local authorities in large villages and small towns in France have responded to such requests and have adopted extremely active investment policies over recent years in order to expand and improve housing stock, to enhance commercial services, and to encourage the installation of light industries that provide additional local employment (Limouzin 1980).

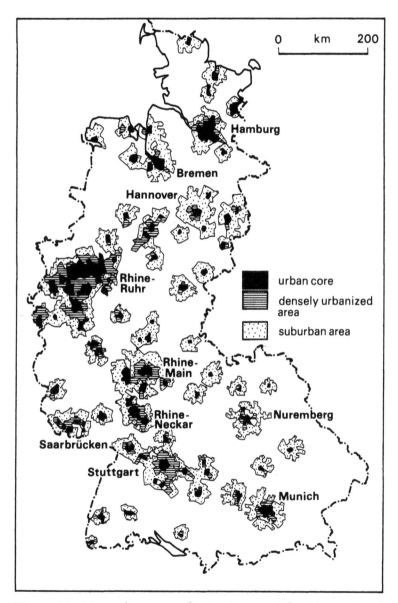

Figure 4.1 Metropolitan areas of West Germany (after Blotevogel and Hommel 1980)

Particularly detailed studies of the advance of the 'dispersed city' and the 'shrinkage' of the countryside have been carried out in West Germany, where changing socio-economic characteristics of local-authority areas have been classified into 'urban', 'urban/rural' and 'fringe' zones that together compose metropolitan functional regions (Wild 1981). Substantial numbers of long-distance commuters do, of course, travel to work from the far countryside that is located beyond these fringe zones. At mid-century 18,450,000 people lived in the fourteen largest metropolitan areas of West Germany that covered 8.6 per cent of the country. Twenty years later the number of metro-politan residents had increased to 29,350,000 (+ 59 per cent) but the amount of space occupied by metropolitan regions had risen by 84 per cent to cover 15.9 per cent of the whole country (Figure 4.1). Fastest rates of population growth were in 'urban/rural' districts close to the 'fringe'. The outer parts of city-focused functional regions have undoubtedly continued to grow substantially since 1970, with suburban land prices continuing to rise, the length of the working day tending to decline (or the pattern of the working week becoming more flexible) and the number of cars in West Germany increasing from 15,000,000 to 23,200,000 during the 1970s. By 1980 metropolitan areas or 'regional cities' with radii extending from 80 km were being recognized around the dozen largest cities in the country (Blotevogel and Hommel 1980). Locations within them, which had formerly contained farming villages and weekend cottages, have been transformed into commuter territory. Exactly the same phenomenon has been identified at comparable distances from most other major cities in the Ten: for example in the mid-1970s property was being purchased and new houses constructed up to 70 km from Central Paris for use as second homes; now such buildings function as the primary residences of daily commuters (Pisani 1979, Belliard and Boyer 1983).

Retirement migration

The residents of the EC are living longer lives, with no less than 38,400,000, namely 5.6 per cent of men and 8.6 per cent of women, being aged 64 years or over in 1980. A growing number of them are choosing to sell their urban home on retirement and migrate to a new location at the seaside or in the country (Warnes 1982). A wide range of interwoven factors explains this trend. Entitlement to old age pensions varies considerably in detail within the Ten but people in each country are eligible to receive financial support in their old age. Greater affluence, longer and more varied holidays, shorter working weeks and rising rates of car ownership have enabled many people to visit a range of places during their working lifetime (Law and Warnes 1980). Retirement offers the possibility of a change of scene and when their children have grown up and left home many couples have no need for their house or flat in town which may be expensive to maintain and on which high levels of local taxation have to be paid. To purchase a smaller property in the countryside or along the coast and use the difference between the two sums to bolster their pensions and savings during their remaining years would appear to be sound economic sense (Cribier 1975).

As with the acquisition of a second home, two types of locational decision are at work. Many people chose to retire to places which they have visited before, perhaps on several occasions, but to which they have no specific attachment. A pleasant climate, attractive scenery and the availability of appropriate and suitably priced housing are the kinds of issue which are evaluated before a decision is ultimately reached. For example, some villages in the mountainous south of the Massif Central retain a range of services and have an appeal among retired migrants, whereas other settlements have experienced so much depopulation and loss of services that the inconvenience of solitude is a very real detractor for inmigration by retired people (Laganier 1977). Other prospective migrants work on a somewhat different set of principles, with the desire to return to the area, perhaps even the specific settlement, from which they originated being the overwhelming factor in their decision-making (Cribier 1982). This is effectively rural-urban migration in reverse and the localities that are repopulated by this second group of retired people may well not even be considered by members of the other group. Return migrants of the second type probably still have relatives and friends of a generation back living in the area who may have kept them informed of local changes and assist them in obtaining a house for their old age. In some instances cottages are reserved or barns converted specially for family members who migrated away in their youth on the clear understanding that they would eventually 'come home'. Throughout their urban exiles some migrants retain possession of plots of land on which retirement houses could be built, while others simply convert second homes into primary residences. Even those returnees who had no specific accommodation awaiting them would probably be able to count on local contacts to help them find a house. At last they were 'returning to the fold' and that fact would condition the kind of reception they might well receive. Family members would be available to supply companionship, help with shopping, and provide support in time of trouble and failing health.

The first group of retired migrants may not fare so well. If they make friends easily or move to a place with a welcoming community atmosphere then the move may not be traumatic, but many potential difficulties await them none the less. They may well have only sampled their chosen settlement or area during the summer months when the sun was shining, shops were open and services were available. They may have virtually no idea of what the place is like in winter when the weather is bad and basic facilities for shopping become sparse out of season. Retirees in small settlements may become seriously disadvantaged as they grow infirm with increasing age. Driving may become impossible for elderly folk who may have to rely on the kindness of neighbours since public transport is sparse or non-existent in many rural retirement areas. The hilly terrain which often contributes to attractive scenery may become a curse rather than a blessing as ageing retired migrants find walking increasingly difficult. Like other residents, retirees support the local economy by making use of village shops and whatever facilities may be available and by paying local taxes, but they also generate new demands for socio-medical services and transport that may not be top priority among other residents (International Labour Office 1971). Unfortunately many rural authorities are simply unable to provide a desirable range of services for elderly people in the countryside.

In short, retirement migration has inflated population levels and in some instances has made a major contribution to halting long-established depopulation in many rural areas in the EC, ranging from villages with favourable climates in Provence or south-west England and settlements in unquestionably beautiful landscapes in Bavaria or the English Lake District, to villages in rather less distinguished districts which none the less have great appeal to returnees. The presence of elderly incomers raises the demand for local shops and can lend valuable support to some village functions, such as the church, but serious difficulties can ensue with advancing years and especially following the death of a partner. Many elderly rural residents need regular domiciliary assistance or even special accommodation but neither is readily available in thinly populated areas with very limited income from local taxes. Some observers argue that a sizeable injection of retired people, with their life savings behind them, produces undesirable competition for housing in villages and small towns and may thereby encourage younger people to move away to find suitable accommodation. Allowing council houses to be purchased by their occupants has introduced other tensions into the housing situation of areas in rural Britain where large numbers of retired people have settled (Dunn, Rawson and Rogers 1981, Gillon 1981). Certainly the local age pyramid becomes more top heavy as a result of inmigration by retired folk and this may dissuade industrialists and others from installing employment for young residents who may have to commute out of the area to work each day, or may decide to move away completely.

The 'Return to Nature'

Even more controversial is the migration of small numbers of people who have chosen to 'drop out' of urban life in order to start farming or handicraft activities in parts of the countryside where housing and land are available cheaply (Allardt 1982). Effectively that means moving to depopulated areas with surplus buildings and where pressures from other incomers are not strong and prices have not been inflated. Doubtless all ten countries in the EC have experienced this trend to some extent but it has attracted particular attention in France, with novice farmers and amateur artisans installing themselves in remote locations, such as parts of the southern fringe of the Massif Central, the Eastern Pyrenees, the mountains of Provence and the Southern Prealps. Cheap property and pleasant climate have had much to do with this choice but during the 1970s novice farmers also settled in parts of north-eastern France and the outermost fringes of the Paris Basin and in harsh environments like the Millevaches plateau of Limousin (Chevalier 1981).

Tiny numbers of city people have 'opted out' for generations but in France the fashion became noticeably more popular during two recent phases. The first occurred after the student troubles of 1968 when a number of utopian communes were set up in the countryside. Few members stayed in them for more than a couple of winters and most of these experiments in communal living disappeared without trace (Léger and Hervieu 1979). A second phase began in the mid-1970s as rates of urban unemployment began to rise. Similar developments also started to occur in Italy, where groups of unemployed and often well-educated young people occupied patches of abandoned farmland. In France the second generation of 'new country-men' were less revolutionary than their predecessors whom local residents simply condemned as hippies. Very often the migrants of the late 1970s were well-educated young couples who found inspiration in ecology and the belief that small could be beautiful. They sought to establish new roots rather than overturn society and in many instances they brought their knowledge and expertise to help local people and to revitalize moribund local activities and village councils. The children of these incomers have been sent to local village schools that often had been threatened with closure. 'New countrymen' have sought to 'get by' rather than generate sizeable profits and the economic basis for doing so includes making pottery, weaving, handicrafts, growing fruit, raising goats or sheep, and making cheese.

The absolute number of such people is minute when compared with the large contingent of French farmers, indeed they probably represent no more than 2 per cent of the rural population of Lozère *département* in the Massif Central and of Ariège in the Pyrenees. But at a much more local scale their presence is important, with perhaps 3000 'new countrymen' living in the Cévennes where they represent one-third or even a half of the population in more remote stretches of countryside (Ardagh 1982). Their babies have been the first local births to be registered for many years in these desperately depopulated areas with elderly resident populations. Local reactions to the early communes, enveloped in rumours of drugs and free love, were unquestionably hostile but the former teachers, engineers and other new ecologists are treated with more tolerance and even a measure of respect, since not only do they bring new ideas but they also appreciate the qualities of traditional rural life (Dessau 1982). In fact their views are often very similar to those held by practitioners of the new, less technocratic brand of countryside planning (see Chapter 8). Mixed reactions have surrounded the arrival of incomers to remote rural areas elsewhere in the EC. Thus urban people who decided to 'escape' to the isles of Orkney during the 1970s in many respects were accepted by local residents, but day-to-day conflicts produced increasing polarization and generated uneasiness about the long-term demographic and cultural implications of the presence of their new neighbours (Forsythe 1980).

Counterurbanization

The combined impact of countryward migration by working households and by retired people has contributed to remarkable changes in spatial patterns of population growth and decline in the EC in recent years (Hall 1980, Hall and Hay 1980). The chronology of these changes varied between countries and regions but census results reveal very similar trends at work. After protracted phases of growth inner-city districts and old-established suburbs are losing residents, while emphatic expansion is taking place in rural areas that have long been characterized by population loss. For example, between 1971 and 1981 the population of England rose by a mere 0.4 per cent, Scotland recorded a loss of 2.1 per cent, while Wales increased its total by 2.2 per cent (Roberts and Randolph 1983). Every large city in Britain lost population, with the largest absolute losses being from inner London (a decline of 535,150 people), outer London (–221,150) and Glasgow (–219,150). The greater proportional loss was 26 per cent from Kensington and Chelsea and every London borough lost more than 10 per cent of its residents in a decade. By contrast, growth occurred along the outer fringes of all metropolitan regions during the 1970s and especially in Eastern England, the south-west and the Anglo-Welsh borderland.

In Italy less urbanized areas also experienced the most important rates of population growth in the past fifteen years. During the 1950s and most of the 1960s urban-rural migration fuelled urban and suburban growth, with many departure areas undergoing serious depopulation (Dematteis 1982). By the late 1960s those trends had begun to change and did so more emphatically in the 1970s (Patella 1981). Inner cities lost jobs and residents, fringe locations gained many incomers, and rural parts of the Mezzogiorno retained people for two quite different reasons. Potential migrants who, under more buoyant employment conditions, would have migrated to urban job opportunities remained in their home areas; in addition, industrialization schemes in the South employed workers who commuted from their home villages each day. Extensive areas on the fringes of all Italian metropolitan regions witnessed dramatic changes in their evolution, with long-established net outmigration and demographic decline being replaced by population growth during the late 1960s and 1970s (Figure 4.2).

Many of the densely built-up, hill-top villages of Provence

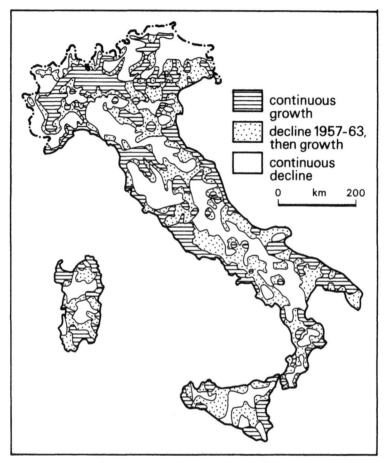

Figure 4.2   Population change in Italy, 1957–79 (after Dematteis 1982)

experienced just the same change in fortune (Joannon 1975). They owed their location, walls, gateways and closely packed structure with sharp angles and narrow passageways to matters of defence that, of course, had become totally irrelevant in the twentieth century. Farmers and other residents had already been moving out of their cramped houses for decades and often established more spacious homes in lowlands or valleys near by. Local services followed their example, with shops, village schools and administrative offices being relocated away from the perched villages where many buildings fell into disrepair. Since the 1950s this tendency for abandonment has changed, with small numbers of artists initially being attracted by the beauty of their site and the possibility of restoring ancient stone-built dwellings that had housed traders and rich craftsmen rather than simple peasants in centuries past (Grosso 1973). Large numbers of commuters, retired folk and other city people have followed fashion and have purchased properties for first or second homes. Assimilation has often been easier for incomers who have family roots in the area. Many have resources to invest in expensive restoration that has generated a great deal of work for local builders and masons and has transformed the appearance of these once-decrepit villages. Water supplies and systems of mains drainage have been installed at great cost. Boutiques and restaurants have opened to serve the new residents and visiting tourists and provide these settlements with a new middle-class vitality. Not surprisingly, the educated newcomers have often gained control of village affairs and have organized conservation societies and festivals (Chevalier 1981). Many of the very same villages which were condemned as dirty and ugly at the end of the nineteenth century have become highly sought after.

These new trends in population change and the decentralization of employment associated with many of them have come to be known collectively as 'counterurbanization' and a number of studies have shown that this process has now replaced cityward migration and continuous suburbanization in most regions as the dominant force in reshaping the Community's settlement patterns (Fielding 1982). But different countries have reached varying stages along the urbanization/ suburbanization/deurbanization spectrum, with Belgium displaying the most pronounced characteristics of the dispersed city and West Germany, the Netherlands and the UK not far behind (Drewett 1980). The arrival of highly mobile incomers and the installation of new employment opportunities has unquestionably brought new life and new prospects to many parts of the EC's countryside but it would be erroneous to think that such transformation has been spread evenly (Courgeau and Lefebvre 1982). Detailed inspection of recent changes reveals plenty of stretches of deep countryside that have little attraction for inmigrants, either employed or retired, and continue to experience depopulation as before. In addition, many individual settlements in otherwise buoyant areas have failed to share in the general trend of inmigration, while profound contrasts in access to housing and other goods and services exist among different social groups in areas of revitalized countryside.

Second homes

As well as the processes that are repopulating many areas with urban-rural migrants, two closely related trends are bringing large numbers of city people into Western Europe's countryside on a temporary basis, namely the use of second homes and the development of various forms of rural tourism. Second homes were occupied in the past by small numbers of affluent city dwellers, and the country retreats of Londoners, Parisians and citizens of provincial France were described by Daniel Defoe and Arthur Young during the eighteenth century. Country villas had also multiplied in Tuscany, Umbria, Venezia and doubtless many other areas of Western Europe (Desplanques 1973). Increasing but still small numbers of country houses were constructed or purchased by the upper and middle classes of urbanizing Europe during the nineteenth centuries and families, often of quite modest means, which had migrated from countryside to town, sometimes inherited farm property that could be used for holidays or at weekends (Clout 1976a and b). The fundamental features of this trend were certainly in existence by the 1940s but numbers of second homes were very limited in all parts of Western Europe except France, which had c. 200,000 rural second homes in 1946. As a general rule, most farmhouses were still needed by people working in agriculture; car ownership was very restricted; and few Europeans had long holidays with pay. Each of those conditions subsequently changed and what may be thought of as the 'second-home boom' began gently in the 1950s and developed with increasing rapidity until the early 1970s when conditions stabilized for a few years (Coppock 1977). Sample enquiries suggest that numbers continued to increase subsequently, albeit at a more modest rate.

Official definitions of second homes vary between countries and special censuses and surveys have not been undertaken in all regions of all member states of the EC. However, scrutiny of available information allows one to estimate a grand total in excess of 3,000,000 in rural areas of the EC in the early 1980s, with no less than 1,400,000 being found in France (Chevalier 1981). The significance of that figure is highlighted by the fact that France now contains only just over one million farmholdings; and despite a small national total and rate of occupation there are certainly also more second homes than farms in the UK. Roughly one-fifth of all French households have access to a second home and a comparable ratio applies to Denmark (and, indeed, the other Scandinavian countries) but ratios fall very much lower in other member states of the EC. In addition, details vary substantially within individual countries in terms of number, pattern and intensity of use during the year, and mode of acquiring second homes (Cribier 1973).

As with retirement 'houses, some second homes are purchased or specially built and their sites are consciously prospected. Other properties are inherited and their occupants do not undergo this kind of spatial sampling. Some second homes are acquired so that city dwellers can simply 'escape' and their occupants will place great importance on isolation, peace and quiet; but for other people a second home is only a means to enjoy defined leisure pursuits, such as skiing or boating, and in such circumstances very specific locational criteria will have been identified and tested (Thissen 1978). These will probably emphasize rather different and less isolated sites than those favoured by the first group of occupants. Finally, the second home may be acquired with the long-term intention of retiring to the country. Thus, the immediate function as a weekend or holiday house is merely interim, and in such cases attention has been paid (albeit often inadequately) to the local availability of services that will be required later in life.

With the passage of time, patterns of second-home development spread outward from urban centres, as means of transport improved and volume of demand grew. For example, the construction of solid asphalted roads in the mountains of central Italy during the 1950s was soon followed by the purchase and rehabilitation of uninhabited houses for use as second homes and retirement houses by residents of Rome and other major cities (Ruggieri 1972). In addition, newcomers constructed new buildings to meet these objectives and in so doing brought standardized housing styles deep into the countryside. As one might expect, weekend cottages are often closer to first homes than are second homes that are only used during long vacations. However, when patterns of development and change are examined in more detail, the importance must be stressed of axes (along motorways and railways), sectors (around pleasant valleys, across varied terrain, on sunny hillslopes, in areas where depopulation has freed housing for alternative uses) and nodes (around lakes, in settlements with socio-cultural attractions or where advertising has been effective or planning permissive).

The social, economic and cultural implications of second homes in the countryside are open to great controversy, that strengthens as the ratio between first and second homes in any given area becomes closer.

Indeed, second homes outnumber primary residences in some localities as, for example, in the outer parts of the Paris Basin, where demand for weekend cottages is strong, or in sections of the French Alps, where depopulation has been particularly intense (Clout 1969a, De Réparaz 1982a). New sources of tax revenue, additional seasonal jobs and rewarding social contacts may arise from the presence of second-home occupants. Ramshackle cottages and barns may be restored to habitable conditions rather than being allowed to become even more decrepit. Local shops and garages may derive trade from the presence of second-home occupants and this additional income may enable them to continue to operate and thereby benefit year-round residents.

But a number of other features counterbalance all this. Young local couples may be unable to afford inflated house prices and may have to move away. This has proved to be a particularly difficult problem in rural areas that have been designated as national parks or endowed with other forms of 'cherished' status and where increases in house prices because of the arrival of incomers have been particularly intense (Clark 1982a, Rogers 1981). Unless siting policies are enforced and high standards of construction maintained, serious landscape deterioration can easily result in areas where large numbers of new houses are built for use as second homes (Gojceta 1978). For example, there has been great concern about the environmental impact of new second homes in Brittany and the construction of new weekend and holiday houses in the Irish countryside has caused profound changes in settlement patterns, both in areas close to Dublin and along sections of the depopulated Atlantic coast (Cléac'h 1977, McDermott and Horner 1978). As with aspects of rural tourism proper, communities may be split into different camps, with some residents and influential personalities favouring second homes and catering for their requirements, while others remain in strong opposition (Clout 1970). Linguistic and cultural distinctiveness may well be diluted by the temporary presence of outsiders and fear of this process, together with the inflation of house prices, has formed a strong motive for formidable hostility to the acquisition of second homes by English people in many localities in Welsh-speaking Wales.

Rural tourism

Tourism and recreation provide another mechanism whereby urban people and urban values are transported into many types of countryside, ranging from areas within easy access of cities for day or half-day visits, to interesting rural landscapes in which to spend weekends, and mountain villages for enjoying winter sports (Barbichon 1973). As Christaller (1964) explained, the very rurality and peripherality of such areas endows them with attractions for members of an essentially urban society. In addition, the presence of visitors introduces perceptions of space and place that are quite different from those held by the farming population. For example, the hills and valleys of south-east Belgium have long experienced depopulation and in agricultural terms are a 'less-favoured area' (see Chapter 8); however their dissected landscapes, woodlands and meadows have proved particularly attractive to visitors from Flanders and the Netherlands. Since 1970 recreation villages and groups of chalets have been built on sunny hillsides to cater for this demand and, in turn, have given rise to severe criticism from rural conservation groups (Vanlaer 1979).

Members of farming communities tend to appraise land in an eco-logical way, understanding the implications of slope, soil quality, insolation, elevation, road access and location of neighbouring properties for their routine agricultural activities. It is a familial and communal patrimony, which has been the source of the social and productive existence of past generations, rather than simply being an exploitable resource. By contrast, visitors and developers concentrate on the potential of individual sites in their own right and in relation to very specific objectives, rather than in a broad ecological sense (Rambaud 1967, 1969). Thus, steep hillslopes with poor soils and limited grazing potential may be ideal for winter sports or building chalets which command good views, and the value that they acquire may be infinitely greater than that of low-lying meadows that were highly prized by generations of farmers. Visitors will evaluate other components of the countryside in quite a different way from rural residents. Farmers may wish to uproot hedges, plough up pastures, fell lowland woods or plant timber in upland areas, whereas urban visitors will argue the case of landscape protection and environmental conser-vation. Country folk may wish to patch up ramshackle farmhouses to improve their accommodation or favour the development of local mining or manufacturing activities in order to create jobs, but many visitors will stress the need for careful (and that usually means expensive) restoration of vernacular architecture and the desirability of protecting the rural environment.

Some landowners, shopkeepers and influential residents will respond readily to the opportunities for employment and profit making that rural tourism can provide but others will be far more reluctant and will try to adhere to their old values. Thus, the development of tourism in the mountains of Provence literally 'split rural communities in two', while catering for summer visitors in the seaside villages of Western Ireland produced exactly the same division (De Réparaz 1982a, 402; Brody 1974). Despite clashes of attitude, catering for visitors on a modest scale, such as bed-and-breakfast in farmhouses and locally-owned inns can generate additional income to complement ailing agricultural activities and thereby play a role in retaining the rural population on the land (White 1976).

In this basic form of rural tourism, catering for visitors who are touring or hiking represents only a minor diversion from the primary activity of farming, but if demand for services grows, both in volume and intensity, the relationship can change and can challenge traditional dependence on the land (Mallet 1978). This kind of evolution has occurred in many mountain villages where, earlier in the century, visitors were few in number and came to walk and relax during the summer (Kariel and Kariel 1982). Later on the fashion for skiing and winter sports developed, with profound implications for those settlements. Visitors demanded higher standards of accommodation and amenities, which were provided initially with the help of local capital but eventually required external investments (Rambaud 1980). Such was certainly the case when it came to building hotels and installing ski lifts and other equipment. Substantial changes to the landscape result and a proportion of the profits leaks out of the local area and back to the investors.

In addition, important social changes occur. Those who cater for visitors acquire more knowledge of the outside world and devote less time to family and friends and to farming, with distant fields and areas of pasture frequently being abandoned. Their lives become more organized to rigid work routines and increasingly geared to profit making from tourism (Kayser 1980). Over time such changes have wider implications on the area as a whole: church attendance among local people declines; more stress is placed on formal education and training; outmigration may decline, in response to the income and new employment generated by tourism; certainly any traditional sense of co-operation between neighbours is replaced by a rising sense of competition between individuals; and there is growing dependence on supplies and services which have to be obtained outside the local area.

In these and many other ways tourist-orientated sections of the countryside are drawn into the urban-orientated society from which their remoteness had largely sheltered them in the past (Vincent 1980). The Alpine areas of France, Italy and neighbouring countries beyond the EC provide the clearest examples of this kind of evolution but it is found in more modest ways throughout the countryside of Western Europe from Scotland to Sicily (Hugonie 1976).

By virtue of continuing demand and the extra income and new employment that can result, tourism in its many manifestations had considerable appeal to rural policy makers. It has already provided a valuable addition to many local economies and can undoubtedly offer support elsewhere. However the process of catering for visitors is never easy or painless, nor should it be thought of as being suitable to provide the sole source of income for rural areas. Destruction of traditional local attitudes and greater subservience to the trends and tastes of urban society is the price that has to be paid (Billet 1976). And there are plenty of examples of sophisticated rural tourism schemes being financed from outside and serviced by workers brought in from elsewhere, so that local people derive little benefit and are employed only in the most menial tasks that have no appeal for many young people who chose to move away (Rosenberg 1973).

5 Rural land and farming structures

The shrinking countryside

As Wibberley (1960) so rightly stressed, the thinning out of the urban mass of Western Europe, together with its increase in area, has no natural or automatic limit, but how much rural land was converted to urban uses in the EC between 1950 and 1980 is unfortunately not known with certainty. However, according to Best (1979, 1981), the amount of urban land in the Nine increased during the 1960s by just over 1,200,000 ha (+ 13.6 per cent) and when an allowance is made for Greece the total rises to roughly 1,290,000 ha, equalling 0.78 per cent of the surface of the whole Community. Italy, with its economic miracle, massive rural exodus and unsophisticated systems of physical planning, registered a 31 per cent increase in urban land during the

Table 5.1 Increase in urban land, 1961–71

	Urban land 1971 ('000 ha)	Urban as % of total surface 1971	Increase in urban land 1961-71*
Belgium	441	14.6	+ 0.8
Denmark	388	9.2	+ 0.6
France	2,577	4.9	+ 0.5
W. Germany	2,858	11.9	+ 1.5
R. of Ireland	103	1.5	+ 0.2
Italy	1,220	4.2	+ 1.0
Luxembourg	17	6.6	+ 0.8
Netherlands	505	15.0	+ 1.3
UK	1,918	8.0	+ 0.8
The Nine	10,027	6.8	+ 0.8

Source: R. H. Best (1981).   *Note:* * Expressed as percentage of national surface.

1960s but despite that only 4.2 per cent of its land surface was in urban use in 1971 (Table 5.1). West Germany (+ 14.4 per cent) also experienced an above-average rate of increase but all other countries augmented their urban area by less than 13.6 per cent, with increases in the Netherlands, Denmark and Belgium being particularly modest.

As was explained in Chapter 1, comparable data for urban change during the 1970s are not available but a broad impression may be obtained from a scatter of official figures together with interpolations on the basis of experience during the 1960s. Certainly the depressed 1970s were very different from the buoyant preceding decade and rates of house building decelerated; however the construction of motorways, peri-urban shopping centres and other large space-consuming projects continued none the less. Working in this way, *faute de mieux*, it would seem that a total area a fraction larger than the island of Sicily was converted to urban uses throughout the Ten during the 1960s and 1970s, and if one were to consider the thirty years since mid-century it is likely that the area would be greater than the total surface of Belgium. It is probable that less than 8 per cent of all land in the EC was devoted to urban uses in 1981 but proportions were high in the Netherlands (16.3 per cent), Belgium (15.5 per cent) and West Germany (13.2 per cent), and above average in Denmark (10.0 per cent) and the UK (9.0 per cent) (Table 1.1). The remaining nine-tenths of the EC was composed of woodland, covering one-fifth of the total area, various types of farmland (arable, permanent grass, permanent crops) occupying almost two-thirds, and finally areas devoted to 'other miscellaneous' uses. One may legitimately echo Best's (1979) words: 'it is still patently obvious that no EC country can even remotely be described as largely covered with bricks and concrete' (p. 403).

Reactions to this evidence range from optimistic acceptance to pessimistic horror. Some would argue that to convert an area the size of Belgium to urban uses is a reasonable price to pay in order to house 48,500,000 additional West Europeans since 1950 and to accommodate a total of 270,857,000 people at higher living standards than ever before. National systems of physical planning differ profoundly and have met with varying degrees of success in containing urban Europe and shaping its permitted growth. Optimists would agree that urban sprawl has been avoided in recent decades in much of north-western Europe, while acknowledging that the same cannot be claimed for Italy and other parts of the Mediterranean South where badly sited industry and housing has been allowed to spill chaotically into the countryside. For example, the rich agrarian landscapes known as 'green Umbria' in central Italy have almost entirely disappeared because of the disorderly expansion of settlement and factories across fertile farmland (De Angelis and Patella 1978). Similar examples of urban sprawl affect the environs of Marseilles, Athens, indeed most large Mediterranean cities (Bonnier and Coste 1978, Castela 1976).

Other commentators would take the same information on land conversion and interpret it the reverse way, emphasizing the 'sacrifice' of an area the size of Belgium and seeing this as clear evidence of the wholesale failure of physical planning. They would stress four main points, namely: the general waste of rural land; the loss through suburbanization of high-quality farmland in the carefully cultivated hinterlands of many great cities; the growth of derelict, unused land in many inner-city localities, while rural space continues to be consumed; and the threats that new housing and tourism projects bring to more remote and relatively unspoiled parts of the countryside. Arguments involving advances in human welfare, the undesirability of fossilizing landscapes, the impact of agricultural intensification and the generation of food surpluses may be introduced with varying degrees of success to counterpoise the first two points, but the problems of inner-city dereliction and threats to especially attractive countrysides cannot be denied; nor can the chaos that surrounds so many Mediterranean cities. The challenge is surely not to halt urban expansion into the countryside until inner-city areas are revitalized but rather to implement efficient systems of physical planning which operate with sensitive knowledge of the ecological, economic, social and political implications of various degrees of restrictive intervention in the process of land conversion (Rogers 1978).

Woodland

Turning away from urban areas into the countryside, it is clear that the Community's woodland experienced contrasting fortunes in the three decades following mid-century. On the one hand, hedgerow timber, individual trees and small woods were felled in great quantity in areas where agricultural intensification, suburban expansion and highway construction were taking place, such as Flanders, the Ile-de-France, the 'low Netherlands' and lowland England. Some 20,000,000 elm trees were lost in the latter region alone during the 1960s and 1970s following the introduction of a new agressive strain of Dutch elm disease through certain English ports (Grainger 1981). The disease spread rapidly through Southern England killing the vast majority of elms and by the late 1970s had become established not only in several parts of Northern and Western Britain but also in other regions of the Community from Denmark to Italy (Burdekin 1983). Many lowland and suburban landscapes throughout Western Europe have been altered dramatically following the clearance of dead elms (Jones 1981). Additional losses due to natural causes involved many of the hedgerow trees that had been planted in lowland England and elsewhere in northwest Europe during the late eighteenth-early nineteenth century enclosure period but recently reached their physical maturity and started to die (Hardy and Matthews 1977). Inadequate development control exposed many private woodlands around Paris, Rome and other major cities to clearance for piecemeal suburbanization (Kalaora and Pelosse 1977). These and similar changes not only changed – and many would argue impoverished – lowland landscapes in many parts of the EC but they also erased valuable recreational resources for urban dwellers (Kalaora 1978).

In purely quantitative terms all these losses were more than counterbalanced by massive increases in tree cover which took place in depopulating upland regions and in other environments deemed to be 'marginal' for farming. Considerable potential remains for clothing such areas as sections of the broad depopulated countrysides of Western Ireland and smaller abandoned areas in Central Germany with trees (Aalen 1978). Two quite different mechanisms were responsible for these increases. Reafforestation programmes managed by national or regional agencies involved sizeable stretches of terrain and gave rise to forests that had a chance of being cropped economically in the future. However, blanketing wide areas with serried ranks of the same – often introduced – species has proved controversial with respect to aesthetics, ecology and possible recreational use (Peterken and Harding 1975). Conservationists argue the need to ensure a variety of tree species within new forests to enable a diverse fauna to survive and to assist natural regeneration (Carbinier 1978). For instance, the survival of new forests in the French Southern Alps is now threatened because the original method of planting produced dense uniform stands of timber which severely reduce possibilities for natural regeneration unless thinning is undertaken (Douguédroit 1981). After a long period of concentrating on the production of trees almost to the exclusion of other matters, European forestry agencies are broadening their objectives. The Dutch forest service, which was set up to produce timber, now has responsibilities for landscaping, nature conservation and provision for recreation use (Beynon 1979). Some state forests in the Netherlands are reserved for timber production but others are zoned for multiple use, with picnic sites, trails and other public facilities being provided at locations which are not inimical to nature conservation or the quality of the landscape. Similarly, of recent years foresters in France and the UK have paid attention to planting schemes that are compatible with existing landscapes and to management techniques that offer the possibility of using parts of the woodland for recreation (Saurin 1980, Dewailly and Dubois 1977).

The second aspect of reafforestation involved landowners in regions such as the Massif Central and the uplands of Southern Belgium and Central Germany who planted up their tiny plots with trees or simply abandoned them to revert to a tree cover (Christians 1980). Many of these owners were migrants or the children of migrants who lived in distant towns and cities and had no direct interest in their woodlots (Larrere 1978, Bravard 1975). Such anarchic, 'wild' or 'postage stamp' reafforestation is fraught with difficulties, producing chaotic, abandoned landscapes that have little or no appeal to visitors and almost never offer the chance of economic harvesting (Clout 1969b).

The net result of clearance and reafforestation has been for the woodland surface of the EC to increase by one-fifth from 29,000,000 ha at mid-century to 34,870,000 in 1980, when it occupied 21 per cent of the total area. Each country enlarged its tree cover, with increases close to 30 per cent occurring in the UK, the Republic of Ireland and France. The latter country also recorded the greatest absolute advance, with no less than 3,175,000 ha (an area greater than the region of Provence-Côte d'Azur) being newly devoted to woodland. In 1980 over one-quarter of the total area of West Germany and France was under trees, with woodland covering very large sections of Aquitaine, Franche-Comté, Southern Belgium, Venezia and much of Central Germany (Table 5.2, Figure 5.1a). By contrast, the tree cover of the Republic of Ireland and the lowlands of England and the Benelux countries was particularly modest.

Three-fifths of the Community's woodland is in private ownership and in France the figure reaches 72 per cent, having been enhanced substantially in recent decades (Table 5.2). As has been suggested, private woodlands are often very small in extent; for example four-

Table 5.2   Woodland: area, ownership and composition (per cent)

	Proportion of total area	State forest	Communal forest	Private forest	Proportion under softwoods
Belgium	20.1	12	35	53	47
Denmark	11.4	30	11	59	61
France	26.8	10	18	72	30
W. Germany	29.4	31	25	44	70
Greece	22.5	65	12	23	33
R. of Ireland	5.3	75	1	24	?
Italy	20.0	6	34	60	16
Luxembourg	34.6	7	37	56	19
Netherlands	7.1	28	16	56	69
UK	8.6	43	0	57	49
EC	21.0	21	21	58	n.a.

fifths of woodland holdings in West Germany are under 5 ha in size and often comprise two or three quite distinct plots (Niggermann 1980). Communal or public forests account for 21 per cent of the EC's woodland cover and, if well managed, generate important financial resources for the local authorities in which they are located (Dietrich 1976). An identical proportion is made up of state forests that provide particularly effective systems of economic management in West Germany, Northern France, the Netherlands and the UK.

Forests in Mediterranean Europe are particularly vulnerable to damage by fire or unauthorized grazing, regardless of their ownership status. In Greece most woodland is subject to grazing, often in contravention of state forestry regulations. As a result, the timber that is produced tends to have little commercial value and regeneration is often impaired. Few precautions are taken against forest fires and these have caused enormous damage to the quality and extent of the nation's timber resources (Kayser and Thompson 1964). Italian woodlands suffer similar destruction with over 20 per cent of the tree cover in parts of Sicily and Tuscany being consumed by fire in a single decade (Figure 5.2) (Morandini 1978). In France the Southern Alps have undergone extensive reafforestation over the past hundred years by spontaneous regeneration and deliberate replanting while sheep grazing has declined substantially, but every summer the tinder-dry woodlands of Provence-Côte d'Azur are easy prey to fires and an average of 15,000 ha of

Figure 5.1   Aspects of land use and farm size in the European Community, 1980

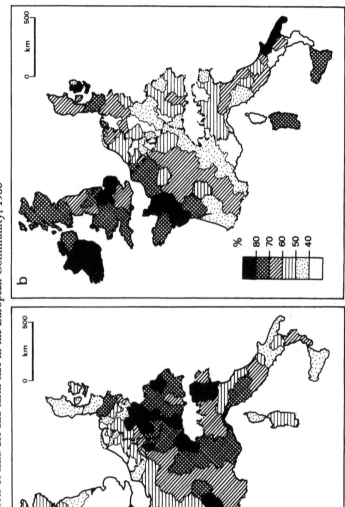

a Woodland as a proportion of the total surface (per cent)
b Farmland as a proportion of the total surface (per cent)

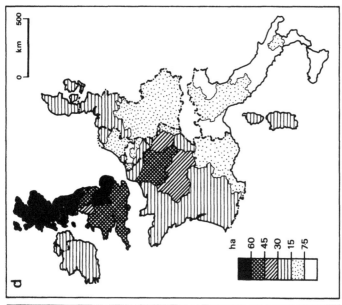

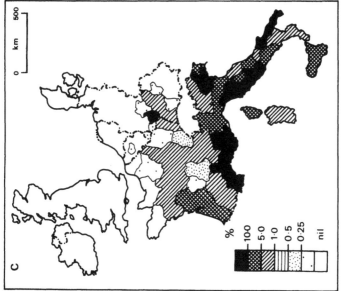

c Vineyards as a proportion of the total surface (per cent)
d Average farm size (ha)

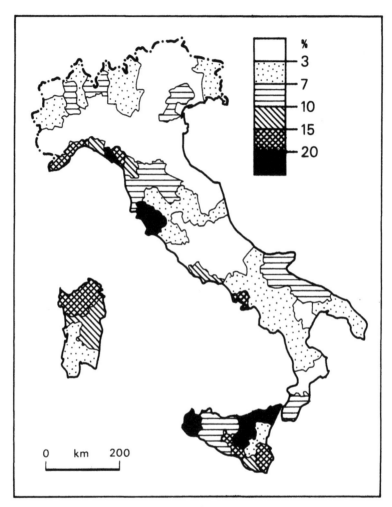

Figure 5.2   Forest areas damaged by fires in Italy, 1964–73 (per cent) (after Morandini 1978)

timber are destroyed each year (De Réparaz 1982b). Despite improved fire-fighting and management techniques the problem seems to be worsening, with growing numbers of campers in and around the forests being an important contributory cause. More attention needs to be directed to educating the public about the vulnerability of what they perceive as 'leisure forests' (Rigole 1972). Further west, the burning of light timber and moorland in order to extend pasture areas is still customary in parts of the Central Pyrenees and the woodland suffers accordingly (Métailié 1978).

Farmland

Some 101,915,000 ha, or 61.5 per cent of the total area of the Community, were in agricultural use in 1980 with national values ranging from 47.4 per cent in Belgium (strongly urbanized, moderately wooded) to 77.5 per cent in the UK (important urban development but relatively little timber) and 81.1 per cent in the Republic of Ireland (limited urbanization and very little woodland). Farmland occupied more than 70 per cent of the land surface of many regions of the British Isles, north-western France, Northern Germany, Eastern Denmark and three sections of the Italian Mezzogiorno (Figure 5.1b). Twenty years earlier the total agricultural surface of the Ten had covered 110,000,000 ha and each country subsequently experienced a notable reduction save the Republic of Ireland where reclamation and reclassification allowed the agricultural surface to increase. In most countries the rate of decline between 1960 and 1980 conformed closely to the EC average (-7.3 per cent) but agricultural land appears to have contracted particularly sharply in West Germany (-11.3 per cent), the Netherlands (-12.4 per cent), Belgium (-13.0 per cent) and Italy (-13.5 per cent). No less than 31 per cent of the Community's total farmland and 42 per cent of its woodland is located in France, the rural 'giant' of the Ten which is roughly equal in size to West Germany and Italy added together.

Just under half of all farmland in the EC (48.2 per cent) is used for producing cereals, roots, sown fodders and other arable crops, with a particularly strong arable bias characterizing land use in Denmark and to a lesser extent in West Germany, France and Italy (Table 5.3). The two latter countries alone contain no less than 57 per cent of all arable land in the EC. At the other extreme, less than one-fifth of Irish farmland is used for field crops and proportions are also well below average in Greece, the UK and the Netherlands. Large percentages of farmland are devoted to permanent grassland, albeit of strikingly varied quality, in four countries (Ireland, the UK, Netherlands and Greece), where the grassland surface greatly exceeds the EC average of 45.8 per cent. Since 1960 the Community's permanent grassland has fluctuated by only a fraction (-0.5 per cent), although there have been

Table 5.3   Agricultural land: area and composition, 1981 (per cent)

	Proportion of total area	Arable	Permanent grass	Permanent crops
Belgium	47.4	51.4	46.0	1.0
Denmark	67.4	90.8	8.7	0.5
France	58.5	54.2	40.5	4.5
W. Germany	49.3	59.3	38.8	1.5
Greece	70.0	32.7	57.0	10.3
R. of Ireland	81.1	17.5	80.7	0.0
Italy	59.3	52.5	28.7	18.5
Luxembourg	50.0	43.8	54.6	1.5
Netherlands	49.2	40.8	57.2	1.8
UK	77.5	36.6	62.9	0.4
EC	61.5	48.2	45.8	5.9

marked increases in Italy (+ 22 per cent) and West Germany (+ 36 per cent) but notable reductions in other countries. The total arable surface has declined by one-tenth, with all member states except the Republic of Ireland sharing in that trend and a remarkable reduction of 25 per cent occurring in Italy.

Permanent crops, including vines, olives and fruit trees, occupy almost all the remaining farmland in the EC. Yielding one of the most characteristic crops of Mediterranean Europe, olive trees play an important role in utilizing southern hillslopes. In addition their extensive root system and resistance to drought enable them to stabilize slopes that could be easy prey to erosion. They cover 1,688,000 ha and are found almost exclusively in Italy (62 per cent of the surface under olives) and Greece. Orchards and fruit bushes occupy a similar total area (1,518,000 ha) but greater emphasis is placed on locations where irrigation is possible. Italy accounts for 58 per cent of that total, far ahead of France (18 per cent) and Greece (12 per cent). But it is the vine that is the most extensive permanent crop, with the landscapes, economies and societies of most Mediterranean areas of the EC bearing its imprint. Thus the vine forms a common characteristic throughout the very varied Italian countryside with over half of the nation's farms producing wine (Tirone 1975). To grow the perennial vine commits a farmer for 30–40 years, requires a labour force with specialist skills and necessitiates the services of specialized establishments (Roudié 1978).

The EC's vineyards now cover 2,709,870 ha which amounts to one-quarter of the world total but they produce 48 per cent of the world's wine (Niederbacher 1982). Roughly 1,000,000 ha yield quality wines, with the remainder producing ordinary table wines. One million farms in the EC make wine for domestic consumption only but viticulture is of commercial importance on a further 2,090,535 holdings, of which 11,400 have over 20 ha of vines apiece, 88,800 have 5–20 ha, and 500,000 between 1 and 5 ha. As one might expect, almost all the Community's vineyards are located in Italy (47 per cent), France (43 per cent) and Greece (6 per cent) but their ownership structure differs considerably, with great emphasis on small family holdings in Greece and Italy, while large estates influence the pattern more markedly in France (Table 5.4). More than one-tenth of all farmland is under vines in Mediterranean France, several sections of Italy, and along the hills overlooking the Rhine and the Moselle in Rheinland-Pfalz (Figure 5.1c). Even Luxembourg has a small but prosperous vine-growing area where viticultural skills are at a premium because of the northerly environment (Sindt 1971).

Twenty years ago the countries of the Ten contained 3,240,000 ha of vines, while a peak of 2,500,000 ha had been reached in France in 1873, Italy had supported 4,500,000 ha at the dawn of the twentieth century, and many regions in other countries are littered with the traces of former vineyards (Muller 1973, Reitel 1973). Viticulture continues to undergo important structural changes, with rural depopulation and abandonment of terraces and hillslopes involving a retreat from upland areas, where vines were often intermixed with other crops, while large modern estates, concentrating on mechanized

Table 5.4   Viticulture in the European Community, 1982

	Surface		Holdings on which viticulture is economically significant
	(ha)	(%)	
Italy	1,281,100	47.0	1,218,630
France	1,164,235	43.0	471,800
Greece	163,330	6.0	309,410
W. Germany	99,930	3.6	89,470
Luxembourg	1,275	0.4	1,225
Total	2,709,870	100.0	2,090,535

monoculture, continue to be planted in lowland areas. Thus, new vineyards have been created since mid-century on the northern fringes of the plain of Languedoc from the Rhône valley to the Pyrenees, while higher traditional vineyards have been abandoned (Carrière 1973). Similarly, viticulture retreated in many upland parts of Italy but new lowland vineyards were being planted at an annual rate of 20,000 ha during the 1970s, especially in the north-east of the country (Tirone 1975). By virtue of their presence on so many family farms, vines (and to a lesser extent olives) are indispensable for the survival of 'typical' Mediterranean landscapes and for helping to keep the agricultural population in employment, with all that implies for maintaining rural services and holding together the fabric of rural society in Southern Europe.

Farm structures

Western Europe's mosaic of agricultural land use is underlain by a complex grid of farm structures that has undergone profound change since mid-century, with the number of holdings that exceed 1 ha falling by two-fifths from 9,776,000 to 5,671,000 in 1980 (Table 5.5).

Table 5.5 Reduction in number of farms of 1 ha and over, 1950–80

	Total numbers		Percentage reduction			
	1950 ('000)	1980 ('000)	1950- 80	1950- 60	1960- 70	1970- 80
Belgium	252	91	64	24	32	30
Denmark	204	116	43	5	26	18
France	2,130	1,135	46	17	20	20
W. Germany	1,648	797	52	16	22	26
Greece	1,000	732	27	n.a.	n.a.	n.a.
R. of Ireland	307	225	27	9	4	17
Italy	3,500	2,192	37	21	21	0
Luxembourg	14	5	64	24	33	28
Netherlands	241	129	47	5	21	29
UK	480	249	48	8	30	20
EC	9,776	5,671	42	17*	21*	14*

Note: *Excluding Greece.

This transformation has operated at different rates and with differing intensities in individual countries. Thus the rate of reduction was particularly great in Belgium and Luxembourg (each -64 per cent) and was notably above average in West Germany (-52 per cent) and France (-46 per cent). On the other hand, farm holdings declined more modestly in Italy (-37 per cent). This trend and the national variations within it has been due to a host of interrelated processes. The expansion of the EC's manufacturing and office-based economies during the 1950s and especially the golden age of the 1960s meant that more attractive and more remunerative sources of employment were available as an alternative to working the land. As a result, many farmers and labourers quit their holdings and moved to new homes in urban areas. Economic recession and mounting urban unemployment in the 1970s and 1980s slowed down both the decline in the number of farm units and the shrinkage of the agricultural labour force (Fennell 1981). A second process contributing to the falling number of farms may be thought of as natural wastage and results from the demise of elderly farmers who either have no family successors or whose heirs are unwilling to work their holdings.

Individual national governments have introduced a wide range of schemes to accelerate the decline in farm numbers and to encourage farm enlargement. Promotion of alternative employment in both town and country is an indirect means of achieving that end, while operation of retraining schemes for agriculturalists and special retirement grants for elderly farmers, who agree to stop working the land and sell off their property, represent direct means. Six member states were already operating agricultural retirement schemes prior to 1972 when they were advocated under an EC Directive (72/160) but even now they have not been introduced in Italy (Calmès 1981, Tracy 1982a). In France, for example, grants have been available to elderly farmers for some twenty years and about 600,000 have chosen to accept them and give up working their land. One-third of all French farmland has been transferred to other farmers in this way. About half of that total has been used to enlarge surrounding properties and the remainder has simply been allocated to new operators with no enlargement taking place. Despite all this activity no revolutionary changes in farm size have taken place in France, indeed there is a complicated system of controls in operation to safeguard the family scale of farming in that country. In addition, it has to be admitted that the economic attractions of the retirement grants have declined sharply in the inflationary years since the early 1970s. In reality, the rate of land transfers in France and other countries of the Community is determined by general fiscal, economic and regional policy, by the laws of landholding and inheritance, and by the level of agricultural price supports rather than by schemes for stimulating the retirement of elderly farmers (Naylor 1982).

Grants may not have been as effective as some had hoped but they have certainly made some contribution to the general downward trend. The Community's farms declined by 1,500,000 during the 1950s and again during the 1960s but then decelerated to a loss of 750,000 during the 1970s. In the latter decade the general tendency was for farms under 20 ha to occupy a smaller proportion of the shrinking general total and for larger units to increase their share. However, the turning point was closer to 10 ha in Belgium, the Netherlands and the Republic of Ireland and, unlike earlier decades, there would seem to have been very little change in farm numbers in Italy.

Despite the 42 per cent reduction in farm numbers in the EC since 1950, profound differences in farm size continue to exist between and within member states. In 1975 the average farm size in the Community was 15 ha and this has subsequently increased to 18 ha (Table 5.6). Only the UK (68.7 ha), Italy (7.4 ha) and Greece (4.3 ha)

Table 5.6   Distribution of farms by size categories (ha), 1980 (per cent)

	Average* size	1.0-4.9	5.0-9.9	10-19.9	20.0-49.9	50>
Belgium	15.4	28.5	19.9	26.7	21.0	4.2
Denmark	25.0	11.1	17.7	26.6	34.8	10.2
France	25.4	20.6	14.5	21.1	30.4	13.3
W. Germany	15.2	32.3	18.7	22.7	22.3	3.9
Greece	4.3	70.9	20.6	6.5	1.7	0.2
R. of Ireland	22.5	14.9	16.7	30.0	29.8	8.7
Italy	7.4	68.5	17.2	8.4	4.2	1.7
Luxembourg	27.6	19.4	10.9	14.5	38.5	16.8
Netherlands	15.6	24.0	20.2	28.9	23.9	2.9
UK	68.7	11.8	12.5	16.0	27.1	32.7
EC	18.0	46.6	17.2	15.0	15.1	6.0

Note: *Average utilized agricultural area per farm.

diverge greatly from the mean, with holdings on the densely-occupied farmlands of West Germany, Belgium and the Netherlands coming slightly below it, and farms in the Republic of Ireland (22.5 ha), Denmark (25.0 ha), France (25.4 ha) and Luxembourg (27.6 ha) exceeding it. But Greece and Italy are characterized by small family farms, with holdings of less than 7.5 ha being the norm over the greater part of Italy (Figure 5.1d) and legislation keeping farm sizes low throughout Greece. Tiny holdings are also typical of Belgian Flanders, with small family farms of 7.5–15 ha being the norm in Southern and Central Germany, much of Benelux, Alsace and south-eastern France and parts of Northern Italy. Only the Paris Basin and the UK have really large average holdings by Community standards. In fact legislation has not only been used to increase farm sizes, since in two countries of the EC it was introduced to produce just the opposite effect through programmes of land reform.

Land reform

In Greece and Italy laws were passed soon after the Second World War to divide large under-used estates and enable the creation of many small land units which could be used to enlarge small holdings or to establish completely new family farms. Important changes in settlement patterns and land use accompanied these reforms which pulverized ownership and ran counter to what was to be the general trend elsewhere in Western Europe. During the nineteenth century most Greeks owned no land but worked on large estates as sharecroppers or paid labourers (Medici 1969). A start was made on land reform between 1907 and 1911 when 106,000 ha were purchased by the state and distributed to over 7000 families. Further legislation starting in 1917 opened the way for much more widespread land reform between the wars. This involved about 60 per cent of all Greek farmland and enabled 180,000 peasant families and 170,000 refugee households to acquire land of their own, while fractionally enlarging a number of existing family farms. The holdings that resulted were certainly small and since the agricultural population continued to grow it was deemed advisable for another phase of land reform to commence. New laws were passed in 1952 allowing 205,500 ha of cultivated land and 145,000 ha of pastures to be expropriated and divided between 135,000 peasant families. The average size of farm created at this stage was 2.0 ha of cultivated land or 22.5 ha of pasture. As a result of inheritance laws and land reform 90 per cent of farms in Greece are owner-occupied and most are fragmented into several plots. Sixty per cent of holdings are under 3 ha apiece with a national average of 4.3 ha and regional figures from 8–10 ha in Macedonia, Thrace and Thessaly to only 2.5 ha on the islands. As part of the land reform programme of the 1950s, 30 ha was set as the maximum size for individual holdings. Recent attempts to promote land consolidation and farm enlargement have met with limited success in this land of émigré property owners.

Despite a long history of land improvement and abortive attempts to legislate for redistributing ownership in the 1920s, Italy's land reform did not commence until 1950. Rural poverty and landlessness were most extreme in the Mezzogiorno and a number of uprisings took place during the second half of the 1940s as soldiers returned from military service with ideas of a more egalitarian social order (McEntire and Agostini 1970). In addition, the Communist party laid heavy emphasis on the need for land reform to wrest control of large under-used estates from their owners and redistribute land among small peasants and landless labourers. Sympathy for Communism was slight in villages that were dominated by the influence of the Church and, of course, among landowners and many professionals, but support was strong in many large villages, and especially in the so-called peasant cities of Apulia where agricultural workers had organized themselves effectively (King 1973). More uprisings occurred as small peasants tried to satisfy their hunger for land and the administration was jolted into action. The reform derives from complicated legislation in 1950 that permitted acquisition and redistribution of land from estates exceeding 300 ha in specified districts that together covered 8,500,000 ha, some 28 per cent of national territory (Figure 5.3). Expropriated landowners were compensated according to the taxable value of their property in 1947 (Dickinson 1954). Newly acquired land was improved (de-stoned, deep ploughed and perhaps drained) and assigned to landless labourers, sharecroppers and small peasants within three years. Beneficiaries were to pay back a proportion of the costs of improvement during thirty years and they were obliged to join reform co-operatives (Franklin 1961).

Two types of holding were created: completely new family farms, usually with a new farmhouse on the property, and small holdings that were to complement existing farms or provide a means of rounding up income derived from other sources. Local reform agencies dealt with

Figure 5.3   Areas affected by land reform in Italy, 1950 (after King 1973)

land improvement, construction of buildings, roads and irrigation works and the establishment of co-operatives as well as providing technical assistance and eduction. Between 1951 and 1962 a total of 680,000 ha were redistributed to 111,500 families, comprising half a million people in all. The most needy members of the community were given priority and hence 44,500 landless labourers received completely new holdings, with farmhouses sited singly or in small groups in the under-used outer ring of land far from the densely-packed southern villages which usually housed 2000–8000 inhabitants, although peasant cities of up to 50,000 existed in Apulia and Sicily. The new farms varied in size according to location and supposed environmental suitability for farming, ranging from 4 ha on irrigated land to 20 ha on poor hill land, around a mean of 7 ha. A further 69,000 small plots were reassigned, either from leasehold to direct property or to enlarge the small existing farms which were far too small to absorb a family's labour.

As in Greece, the Italian land reform created a mass of small holdings that have profoundly influenced the present pattern of farm structures. The major spatial change was to transfer some 200,000 people away from large nucleated villages to new, relatively dispersed farmhouses. In addition, the reform encouraged a shift of farmers from over-populated upland areas to surrounding plains. In some localities, especially those where irrigation allowed intensive cultivation of vines, fruit and vegetables, the reform may be called a success. Thus in the Sele plain, south of Naples, and the Metaponto plain stretching west and south of Taranto many of the reform farms have emerged as viable holdings (McNee 1955). Results were less sure in drier areas where the old *latifondo* economy of wheat and sheep grazing was replaced by more intensive mixed husbandry. Smallness of farm size, absence of irrigation and lack of skill and enterprise among many of the new farmers, who were quite unused to working without supervision, made many of the new farms truly marginal. In yet other localities, such as parts of Sicily and Calabria, the new farms and additional plots failed miserably with poor-quality land being appropriate only for afforestation or for extensive pastoral farming or cereal cultivation. This is what Desplanques (1957) has called the revenge of the physical environment. Infertile land, soil erosion, isolation, inadequate training, dislike of living away from the old nucleated villages and a host of other physical and human causes combined to encourage many families to abandon their new farmsteads and whole reform villages were deserted in some instances. By 1962, 15 per cent of the new farmers had left their holdings and on many of the remainder at least one family member had acquired off-farm employment or had migrated to an urban job, supplying remittances to keep the family 'farm' going. In such cases the reform holdings were simply used as part-time bases in the countryside. More than thirty years after the event some aspects of the land reform may be called disastrous and the whole programme now seems to be a dated and expensive irrelevance in the overall strategy of rural development.

Part-time farming

Official statistics unquestionably provide an incomplete picture of farm structures in the EC since figures are not available on miniature holdings of less than 1 ha nor is there comprehensive information on part-time farming, whereby members of farm-based households are gainfully employed in jobs other than or as well as working the family farm. Indeed many miniature holdings are operated in this way. At first glance, part-time farming may appear to be a relatively simple phenomenon; in reality just the reverse is true since it has a great diversity of origins, labour inputs and income characteristics, and its component subtypes are evolving in varying ways. A special study found that part-time farming was remarkably widespread in the Community and was encountered in one form or another in all regions, regardless of their state socio-economic development (OECD 1977). In the early 1970s part-time farmers formed a high proportion of all farmers in West Germany (55 per cent), Belgium (43 per cent) and Italy (38 per cent), with rather smaller percentages in the Netherlands, France, the UK and the Republic of Ireland (each between 22 and 26 per cent) (Table 5.7). At a conservative estimate there were 2,000,000 part-time farms exceeding 1 ha at that time and if miniature holdings were included the total would have been

Table 5.7 Full-time and part-time farmers (per cent)

| | | Full-time farmers | Part-time farmers | |
			main income from farming	supplementary income from farming
Belgium	1970	56.7	9.1	34.2
France	1970	77.4	5.8	16.8
W. Germany	1965	40.9	25.7	33.4
	1975	45.2	15.3	39.5
Greece	1980	50.0		50.0
R. of Ireland	1972	77.8		22.2
	1982	58.0		42.0
Italy	1970	62.4	5.0	32.6
Netherlands	1975	74.1	6.3	18.4
UK	1973	77.0		23.0
	1978	73.0		27.0

considerably greater. West Germany and Italy each contained just over one-third of the total.

Later enquiries revealed that in 1980 27 per cent of a much smaller total number of farmers in the EC declared that they had a gainful activity outside their farm, of which one-fifth involved work on other farms and the remainder employment in other sectors (Commission of the European Communities 1981, Frank 1983). This generated a total of 1,340,000 part-time farms but once again the total would be substantially higher if one could include holdings where supplementary activities occurred on-farm (e.g. provision of camp grounds, caravan sites, farm guesthouses) and those where the farmer's wife or other family members had off-farm work. Detailed case studies suggest that the number of part-time farmers may well be on the increase; for example between 1971 and 1979 their number as determined from the UK agricultural census rose from 68,000 to 80,000, that is from 23 per cent to 27 per cent of the total (Gasson 1982). In the Republic of Ireland they increased from 22 per cent of all farmers in 1972 to 42 per cent in 1982, while in Greece about half of all farmers now have some other kind of occupation (Cawley 1983). One of the main traits of Italian agriculture during the 1970s was the widespread diffusion of off-farm employment which converted numerous full-time family farms into part-time operations. Experts disagree on the extent of dual-activity farming, with some suggesting that up to three-quarters of all Italian farming families were engaged in obtaining some kind of external income in the late 1970s, but all stress that part-time farming is remarkably widespread and is very likely to be a permanent feature of the Italian countryside (Pieroni 1982). Similar conclusions could doubtless be reached in other parts of the Community.

Part-time farming is certainly not a new feature in the Ten with some versions dating back several centuries or more. There were many historic examples of family-farmers being engaged in hunting, lumbering, fishing, mining, village crafts or provision of local services either throughout the year or at times when agricultural work was slack (Mignon 1971). With the improvement of transport systems, the spread of manufacturing and the development of tourism during the present century a new range of jobs came into existence and were taken on by family farmers who still continued to work their holdings (Clout 1972). They were encouraged to do so by a persistent cost-price squeeze working in tandem with rising income aspirations. These folk became the 'worker-peasants' or 'five-o'clock farmers' who put in a few hours' work on the land once their other job was completed for the day. Sometimes off-farm employment was available locally but very often commuting was involved by public transport, works' bus or increasingly by private vehicles (Figure 5.4). The German village of

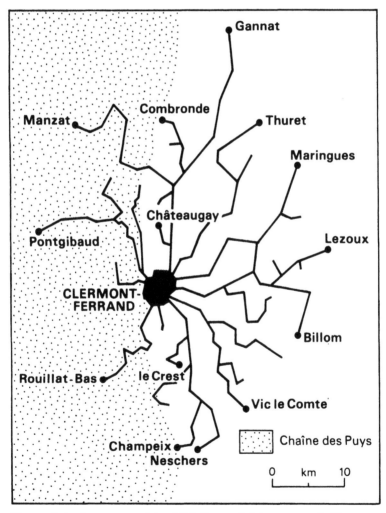

Figure 5.4 Routes served by Michelin works' buses around Clermont-Ferrand (after Bouet and Fel 1983)

Gosheim on the edge of the Swabian Jura provided an example of the first kind of situation, with a range of workshop industries producing nuts, bolts and screws and flourishing on the basis of cheap local supplies of worker-peasant labour from the First World War right through to the economic miracle of the 1960s (Franklin 1964, 1969).

Coalmines, steelworks and other kinds of industrial plant in many parts of Western Europe make use of dual-activity labour, which often commutes considerable distances to and from work each day. For instance, recent industrialization schemes and attendant construction works around Galway City and elsewhere in Western Ireland derive an important proportion of their workers from commuting heads and members of farming families, and in France large concentrations of worker-peasants are to be found in the industrialized valleys of the Northern Alps and in north-eastern parts of the country (Küpper 1969, Cawley 1979, Delbos 1979). In Alsace part-time farmers worked one-fifth of all farmland and represented two-fifths of all farm operators in 1975, with that figure rising to 45 per cent in the valleys of the Vosges (Figure 5.5). The average part-time farm in the whole region was 9.5 ha but most of those down on the plain of Alsace were less than half that size. In West Germany a traditional association between industry and part-time farming is clearly in evidence in the Saarland and Baden-Württemberg where over half the holdings are run by worker-peasants (Mrohs 1982). Other important concentrations are found in more obviously rural areas such as the Bavarian Forest and other middle-range mountain areas where farming is disadvantaged by conditions of soil and climate, dispersed plots (deriving from inheritance customs), small farm sizes, and a lack of local off-farm employment. In such areas up to two-thirds of farms are worked on a part-time basis, with worker-peasants often having to cope with long and expensive commuting journeys each day. No less than one-fifth of the total agricultural surface in West Germany is now worked by dual-activity farmers (Mrohs 1983).

Worker-peasant units, whereby farming people add other forms of employment to their work on the land, are found both in industrialized stretches of countryside and in more distant rural areas. Another type of part-time farming is quite different in character and often in location since it involves urban people choosing to acquire farmland and buildings within relatively easy access of major cities. The quest for space, attractive housing and an alternative residential environment are important factors in motivating this type of part-time

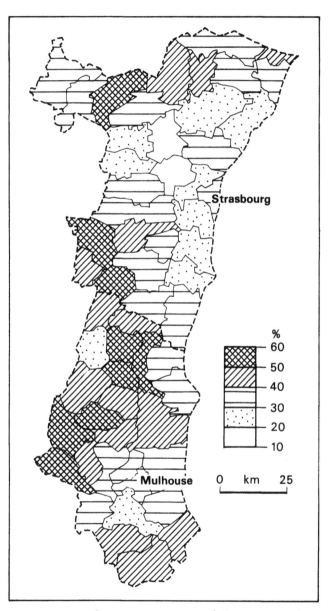

Figure 5.5   Part-time farmers as a proportion of all operators in Alsace, 1975

farmer who has capital to use and is well aware of the financial advantages of investing in land (Gasson 1966). Their holdings are often described as 'hobby farms', since running them is secondary to their owners' urban employment, but the term should not be taken to imply inefficient operations (Barberis 1973). In many instances just the reverse is true. Indeed, Italian experts stress that some 'professional' hobby farms, generally located on plains or in low hilly areas and surrounded by commercial holdings, are run in an exceptionally rational manner and are more productive, receptive of innovations and more competitive than many full-time family farms (Pieroni 1982). But it must be admitted that the quality of management on part-time farms varies enormously.

By virtue of the diverse nature of dual-activity farming it is hardly surprising that evidence from different areas is not in full agreement. Some types have derived from conditions of marked personal hardship, with the family farmer taking on additional work in an attempt to raise his family's standard of living being the classic case. By contrast, hobby farms are the products of well-being and money to spare. Many observers in the 1950s and 1960s believed that the worker-peasantry formed a transitional feature which would soon disappear in Western Europe. They noted that many worker-peasants and their family members lived particularly hard lives as they tried to cope with work on and off the farm and had to put up with lengthy commuting each day. Dual jobholding obviously caused much stress and strain, in some instances affecting the health of the working couple, and might be to the detriment of raising children. In many countries part-time operators were not entitled to benefits that were available to full-time agriculturalists and sometimes experienced difficulties over pension rights and sickness or unemployment payments. Part-time farmers were sometimes moved to encourage their children to migrate to town on a permanent basis. Hence it was argued that being a worker-peasant was merely a phase between running a farm full-time and leaving the land for good. Doubtless that kind of trend did occur in thousands of cases but there is also evidence to show that worker-peasant households can be remarkably resilient, preferring to continue working the land and enjoying the security of living in their own farmhouse as the threat of losing urban/industrial employment loomed ever more ominously after the mid-1970s. The opportunity to work in the open air and experience a more 'natural' environment represented another supporting factor. In addition, the proliferation of hobby farms has generated a new injection of part-time operators and that adds another explanation as to why the 'transition' hypothesis is not as clear cut as many observers believed a decade or so back. The whole matter of dual-activity farming has proved to be remarkably contentious. On the positive side the symbiosis of farm-based and off-farm work provides enhanced incomes for the families involved, although they may experience considerable stress and strain. In addition, the possibility of being able to return to working the land full-time offers an important element of security in an increasingly uncertain world in which industrial jobs are being lost by the million. (Exactly this kind of trend to return to the land occurred in many areas of Western Europe during the depressed 1930s.) Part-time farmers represent a significant component of the total population in many rural areas and their presence maintains a demand for shops, schools and other services which are to the benefit of local residents as a whole. In some instances, part-time farmers provide useful accommodation for visitors, in the form of bed-and-breakfast facilities and camp or caravan sites (Bouquet 1982). For example, just under one-tenth of total tourist accommodation capacity in West Germany is provided on farms (Figure 5.6). Many part-time farmers make significant contributions toward maintaining the fabric of the countryside through simply working the land and using part of their off-farm income to keep farmhouses in decent repair (Caron 1975). By managing land resources they play an especially important role in mountain areas, thereby helping to control floods, prevent landslides and avalanches, and conserve landscapes and water resources (Fuguitt 1977).

On the other hand, part-time farming displays a number of more questionable features. Dual-activity holdings vary considerably in size and specialization, ranging from pastoral farming to viticulture, but they are all run with limited inputs of labour and this fact often restricts the scope and specialization of manageable activities (De Réparaz and Bernard 1975, Auriac and Bernard 1972). Part-time farms are often fragmented and are smaller than full-time units; for example, the average full-time farm in West Germany is 22 ha but its part-time counterpart is only 5 ha (Mrohs 1982). Neighbouring full-time farmers complain that the presence of part-timers blocks their own plans for farm enlargement and rationalization. Indeed not all dual-activity holdings are neat and tidy and in some instances they may be abandoned almost completely, with large kitchen gardens surrounding

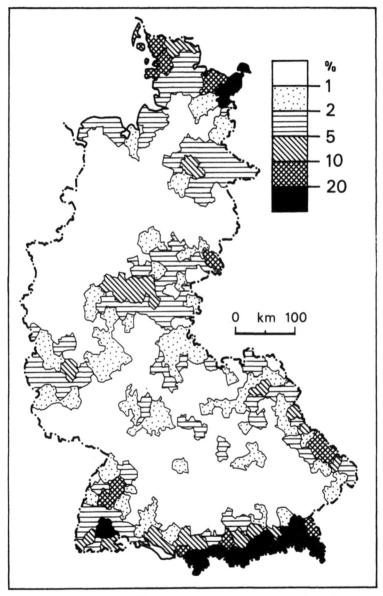

Figure 5.6   Proportion of farms letting rooms to visitors in West Germany, 1975 (after Ruppert 1980)

dwelling houses being the only piece of 'farmland' to survive. This kind of phenomenon is known as 'social fallow' and derives from reasons internal to the farming family rather than because of poor soil or physical isolation. Early studies stressed the importance of social fallow among farms in south-west Germany whose owners enjoyed considerable prosperity as they acquired off-farm work (Hartke 1956). As labour inputs are reduced for one reason or another, so plots of farmland degenerate into scrub or rough woodland and become a source of weeds and pests to plague surrounding farmers. In parts of Northern Lorraine, where many part-time farmers commute to jobs in factories and mines, up to one-third of the land surface has been left as social fallow in some localities (Brunet 1974). Similarly almost half of the farmland in the environs of Rome was abandoned totally or partially during the late 1960s and attractive and well-tended landscapes became severely degraded (Sermonti 1968). For all these reasons social fallow is a source of profound aggravation to full-time farmers whose attempts to improve or enlarge their holdings are foiled by five-o'clock farmers who cling on to their plots as insurance against an unknown economic future. However, attitudes regarding social fallow are changing among some urban dwellers who are starting to appreciate its ecological variety and ability to support a high intensity of bird and animal life by contrast with the monotonous functional environments that have been produced by modern forms of agricultural organization. Indeed, some long-established tracts of social fallow in West Germany have been designated as nature conservation areas, and land that was once perceived as a problem is now being accepted as an important environmental resource (Wild 1983).

Despite their numerical importance, part-time operators occupy a relatively small share of the total farmland in the EC and generate a modest proportion of total agricultural output. For these reasons most governments paid them scant attention and that attitude was bolstered by powerful lobbying from full-time farmers who insisted that dual-activity operators were no more than nuisances. In the densely populated Netherlands, where pressures on land for full-time farming, recreation and urban uses are unquestionably great, the authorities certainly viewed part-time farmers unfavourably and excluded them from EC assistance for the improvement of farm structures (Laurent 1982). They were at a disadvantage when land consolidation took place, since they would be the last to be allocated new fields and could even have them assigned to land reserved for later acquisition by full-time farmers or other eventual uses. By contrast, the West German government adopted a wide definition of agricultural policy in 1968 in order to try to serve the interests of all those who worked in rural areas, who sought their recreation there, and who derived food supplies from local agriculture. To enhance living conditions for all rural dwellers became a central policy objective and the utility of part-time farming as a mechanism for raising incomes and – in more sinister vein – for coping with industrial recession was stressed. Under any circumstances the presence of such a large contingent of part-time operators (and electors) could hardly be ignored. Subsequent rural policy in West Germany has contained a number of original features which encourage the further development of dual-activity farming where the non-agricultural element is dominant, and enable part-time farmers to obtain various forms of official support and advice which had only been available to full-time operators.

Attitudes in France have changed to some extent in recent years and now recognize the worth of part-time farming in difficult environments where depopulation and extensive land abandonment threaten. In response to EC policy enunciated in 1975, a special allowance started to be paid to full-time farmers in upland areas of France in order to compensate for natural handicaps and in recognition of their role in maintaining the cultural landscape (see Chapter 8). Part-time farmers in the Alps, Vosges and other regions where they were numerous complained bitterly that they were being discriminated against and lively debates ensued on the merits and disadvantages of their activities. While being careful to respect the sensibilities of full-time operators, the French authorities now accept that part-time farmers have important roles to play in 'marginal' agricultural areas and have extended various forms of assistance to them. The same kind of reappraisal is finding favour in the European Parliament and other institutions and deserves even wider recognition in the light of changing conceptions of work and leisure, employment and unemployment in post-industrial societies.

6 Transforming the farming environment

An agricultural revolution

The countries of the EC have shared in a veritable 'agricultural revolution' since mid-century that has produced profound changes in the character and scale of farming, has drastically transformed many farmed landscapes, and has had serious implications for the ecology of the countryside (Shoard 1981). Large supplies of labour moved out of the farming sector and have been replaced by unprecedented injections of capital that have been used to acquire machinery, chemicals and other means of intensifying food production. At the same time, an increasing proportion of farmers have been made aware of changing commercial demands and have drastically reorganized their technologies, patterns of production and degree of market orientation in order to satisfy them. Government agencies, co-operatives, marketing and processing corporations, and many other institutions have performed fundamental roles in generating and sustaining this transformation which has been fuelled by finances deriving initially from individual countries and latterly from the costly budget of the Community's Common Agricultural Policy, all set in the context of changing international agreements on trade and aid.

In the spirit of the Treaty of Rome, the CAP operates to supply food to the Community at reasonable prices; to provide a fair income to farmers on well-run enterprises; to help enlarge farms and improve their management; and to be sensitive to the needs of trading in food products with the rest of the world. Commercial farmers throughout the EC have been duly drawn into a new web of relationships that has made them increasingly dependent on critical decisions taken by supranational institutions and by powerful multinational trading corporations (Jansen 1969). Long-established man/land relationships, mediated through farming practices, have been swept away and replaced by intensive systems of great productive potential but also endowed with enormous power for environmental damage (Diry 1975). Rather than respecting traditional countrymen's lore, modern agriculture works to enhance those systems and thereby increase the yields of a few chosen species of plants and animals. Such practice is quite contrary to the basic principles of ecological diversity and the fundamental objectives of environmental conservation.

The net result of this controversial revolution in agriculture has been a reorientation of many aspects of production and a vast increase in farm output from an agricultural surface that has diminished by less than 10 per cent since mid-century. The countries of the Ten more than doubled their output of wheat from 23,000,000 metric tons in 1950 to 51,400,000 metric tons thirty years later and average yields also rose massively from 18,700 kg/ha to 43,600 kg/ha (+135 per cent). National yields now exceed 50,000 kg/ha in the Netherlands, the UK, France and the Republic of Ireland, with West Germany, Belgium and Denmark not far behind (Table 6.1). They are substantially lower in Italy and Greece and it is the fact that Italy is the second largest wheat producer in the EC that serves to dampen down the Community's average yield. Of all ten countries, France has made the most remarkable progress, managing to increase yields by 175 per cent and raise her share of EC wheat production from 33 per cent in 1950 to 42 per cent thirty years later. Community output of barley

Table 6.1 Aspects of agricultural mechanization and productivity, 1980

	Tractor/farm ratio*	Combine/farm ratio*	Wheat yields 100 kg/ha	Potato yields 100 kg/ha
Belgium	1.1	0.1	46.9	306.6
Denmark	1.6	0.3	46.8	250.3
France	1.3	0.1	51.7	289.3
W. Germany	1.8	0.2	48.9	259.4
R. of Ireland	0.5	0.02	50.5	242.6
Italy	0.4	0.01	26.9	181.2
Luxembourg	1.8	0.4	30.8	300.3
Netherlands	1.3	0.05	62.0	364.1
UK	1.8	0.2	58.8	344.8
The Nine	1.0	0.1	43.6	279.7

Note: *Expressed as 1 : x and normally rounded to the nearest whole number.

(+ 430 per cent) and green maize (+ 560 per cent) has also increased phenomenally, with France once again being the largest single producer of both commodities. By contrast, rye production has decreased by one-third in the EC and the output of potatoes has fallen by almost one half from 66,500,000 metric tons to 32,250,000 metric tons, although average potato yields have risen from 180,000 kg/ha to 279,700 kg/ha, with the intensive farming systems of the Netherlands topping the productivity league at 364,100 kg/ha.

In the realm of livestock farming, sheep numbers rose modestly (+ 24 per cent) between 1950 and 1980, as did cattle (+ 36 per cent) but much more dramatic increases occurred in the case of pigs (+ 151 per cent) (Table 2.3). Sheep numbers remained fairly stable or even declined in every country except France, while the most dramatic increases in numbers of cattle (+ 89 per cent) and pigs (+ 543 per cent) occurred in the Netherlands, where a wide range of concentrates and supplementary fodders are used to maintain exceptionally high stocking densities. In some parts of the country this produces large surpluses of manure which are drained, stored and eventually dumped during those periods of the year when demand for manure is low. Such practices give rise to pronounced chemical enrichment (i.e. eutrophication) of surface water and ultimately of ground water as well, which has serious implications for the local environment (Tellegen 1981). Dutch productivity of winter wheat, potatoes, and sugar beet is the highest in the world and with such a striking record of livestock raising it comes as no surprise to learn that farmland in the Netherlands receives three times the EC average application of nitrogenous fertilizer. Similar intensities characterize Belgian Flanders, where 'industrial' farming methods have been adopted widely; however, less intensive production in Southern Belgium serves to depress national figures below those for the Netherlands (Christians 1980). In short, the Community has massively increased its farm output over the past thirty years and has become self-sufficient in cereals and has almost achieved that distinction for meat and vegetables (Table 6.2). 'Mountains' of surplus wheat and butter and 'lakes' of surplus wine and milk are less laudatory manifestations of the CAP and critical conclusions must also be drawn with respect to the impact of highly subsidized intensive farming on the ecological balance. Indeed many would argue that modern agriculture, rather than the relentless march of urbanization, is the main threat to wildlife, the landscape and the whole rural environment of the Community (Green 1975, 1980).

Table 6.2   Aspects of agricultural self-sufficiency, 1980*

	Wheat	All grains	Vegetables	Fresh fruit	Citrus fruit	Wine	Cheese	Butter	All meat
Belgium & Luxembourg	86	57	115	61	0	7	39	112	121
Denmark	120	112	70	53	0	0	451	209	315
France	196	168	94	96	3	114	114	122	97
W. Germany	104	87	34	53	0	53	93	132	89
Greece	135	87	108	163	126	115	92	77	80
R. of Ireland	47	82	88	22	0	0	544	295	278
Italy	80	69	124	131	115	143	80	68	99
Netherlands	58	26	186	47	0	0	225	255	210
UK	77	80	74	32	0	0	72	51	77
EC	112	98	97	83	43	112	107	120	99

Note: *100 = self-sufficiency.

The number of farms in the EC has fallen by half since 1950 and the number of workers by two-thirds, but the new agricultural revolution has necessitated marked increases in the degree of multi-purpose and specialized mechanization. The Community's tractor pool rose from 1,342,500 at mid-century to 5,200,000 in 1980, which almost equalled the number of farms of more than 1 ha at that time. As one might expect from the evidence of 1950, tractor ownership was particularly pronounced in the northern countries of the Community (West Germany, the UK, Denmark), while the ratio between tractors and farms was particularly low in Greece and Italy (Table 6.1). The number of combine harvesters rose from a mere 55,200 in 1950 to 476,000 in 1980, with the highest densities per 1000 ha of cereal land being found in West Germany (32), the Netherlands (28) and Denmark (23) where farm sizes were moderate by Community standards. Not surprisingly, combines remained rare on the small farms of Southern Europe and in the strongly pastoral environment of the Republic of Ireland. The distribution of the EC's 1,310,000 milking machines also reflects variations in farm structures (in this case herd size) as well as trends of diffusion and adoption. For the Community as a whole there were 46 milking machines per thousand dairy cows in 1980, with figures ranging from 85 per thousand on the well-mechanized, moderately sized holdings of West Germany to 29 per thousand in the Republic of Ireland, where levels of mechanization remained low (Table 6.1).

The acceptance of the new agriculture, with machines replacing human and animal power and chemicals replacing natural fertilizers, has changed the chronology of the farming year, has wrought profound changes in the rural landscape, and has had dire implications for the local ecology. Writing of West Germany at mid-century, Niggermann (1980) recalled that the harvest period lasted several months during which farming landscapes were characterized by hay on tripods, heaps of clover and other fodder, and sheaves of grain. Now it takes the pick-up press only a few hours to process hay from the raw state to bales. A combine can harvest the grain of all the farms in an entire village in just a few days. Harvested fields are straight away worked over with multiple ploughs and they take on the appearance of autumn at the height of summer. Mechanization, coupled with farm enlargement, encouraged substantial internal changes on many holdings throughout the EC and as larger pieces of equipment came on the market so the demand to remove field boundaries, trees and old trackways became greater. Patches of ill-drained, low-lying ground were seen as impediments to further progress and were duly drained with financial assistance from the authorities. On many farms long-established crop rotations were abandoned, since application of artificial fertilizers could virtually guarantee a good yield of the selected product. Widening branches of livestock husbandry no longer depended on the natural growth of grass, since animals could be fed on fodder concentrates and their progress controlled according to factory-like routines.

In Southern Europe innovative irrigation techniques, at both local and regional scales, enabled wider ranges of fruit, vegetables, fodders and other crops to be grown in larger quantity and with greater certainty than ever before. Large new schemes have created completely new productive landscapes in many Mediterranean parts of the Community. For example, sizeable areas in the lower Rhône valley and sections of the neighbouring lateral valleys (e.g. the Ardèche, Drôme and especially the Durance) have been planted with orchards producing apples, apricots, pears and peaches following the provision of irrigation water from the Compagnie Nationale du Rhône and other corporations (Brunet 1974). Likewise, the coastal plain of Languedoc has undergone spectacular changes since 1960 with traditional openfields being replaced by regularly-shaped irrigated fields and ancient villages being complemented by new isolated farmsteads (Carrière 1975). The landscapes of the back-country have also been changed, with new vineyards contouring the shape of the hills and some areas of difficult terrain falling out of cultivation. In addition, the use of small motor pumps and plastic or metal piping has enabled small pockets of land to be watered efficiently in the difficult agricultural terrain of Greece and elsewhere in the Mediterranean fringe of the EC (Burgel 1975). In total, the quantity of irrigated land in France increased by 25,000 ha each year during the 1970s and in Italy irrigation involved a further 60,000 ha in the 1960s and over 150,000 ha in the 1970s as a result of the activities of the Cassa per il Mezzogiorno and other agencies. No less than 3,500,000 ha of Italian farmland are subject to irrigation (one-fifth of the total) and the surface continues to increase with assistance from the farm fund of the EC (Bethemont and Pelletier 1983). Similarly in Greece the irrigated surface more than doubled in the two decades following 1960 to reach 910,000 ha in 1980, with the current target being 1,600,000 ha. However, other aspects of agricultural modernization have produced even more dramatic effects.

Land consolidation

The revolution in technology and organization that has permeated so much of the Community's farming in recent decades has transformed many traditional rural landscapes, especially in regions where agricultural intensification has been widely accepted. Field boundaries, isolated trees, copses, small patches of roughland and other micro-features have become particularly vulnerable and have been removed in great quantities. The precise combination of circumstances generating these changes has varied from area to area and from time to time. In some instances they are due to decisions taken by individual farmers about how to run their holdings and claim grants and subsidies on offer from government agencies. But in other cases they are the result of land reallocation schemes which operate over much more extensive areas. Indeed a number of countries in the EC have implemented special programmes for reorganizing fragmented patterns of landownership and for creating consolidated holdings. In some instances their objectives are considerably wider than this and amount to little short of integrated rural development plans. They all produce profound changes in the rural landscape and have frequently been at the centre of controversy because of the ensuing ecological damage. Historic enclosure schemes transformed landownership patterns and rural landscapes in parts of the British Isles, Denmark, North Germany and some other locations, but after the Second World War the fragmentation of landholding into a vast number of tiny strips or blocks intermixed with property in the possession of other landowners remained an important characteristic of land occupation throughout much of continental Western Europe (Lambert 1963). It is arguable that this kind of fragmentation offered several advantages to farmers in earlier decades when agricultural tasks were predominantly undertaken by hand. First, each owner of a number of scattered plots could make use of various types of soil and terrain, thus permitting a range of land uses to be accommodated and different types of crop to be produced. This was clearly beneficial in the kind of mixed farming, semi-subsistence economy which had operated in most areas before railways and main roads facilitated the exchange of goods between regions. Indeed, such systems of production were still to be encountered in the 1940s in parts of Mediterranean Europe and in isolated localities elsewhere in the Ten. Second, property fragmentation served as a kind of insurance against agricultural disaster, since a scatter of plots increased an owner's chance of escaping localized hazards such as drought, flood, frost and hail. Third, small parcels were appropriate for cultivating by hand; and, finally, a fragmented pattern enabled small owners and tenants to enhance their social status by gradual purchase of small plots as their savings allowed.

In fact the disadvantages of farming strips of land in areas of fragmented ownership were becoming obvious after 1945 as agricultural labour started to become scarce and mechanization began to be adopted. Operating costs were inflated because of the time farmers had to devote to travelling between dispersed plots and perhaps also transporting machinery and other equipment (Chisholm 1962). Potentially useful farmland was sacrificed in areas where property divisions were marked by a proliferation of ditches, banks or hedges. Ecological problems included shading of crops, interrupted drainage and invasion by pests from badly farmed adjacent plots, while in some localities strips were orientated without due regard to slopes and soil erosion resulted. In addition, proprietors of plots that lacked farm tracks and were surrounded entirely by neighbours' land had to contend with serious problems of access. Complicated agreements had to be reached on this topic and on the type and chronology of farming activity that would be tolerable to all parties involved. Farmers might find their plots too small for mechanized working and start abandoning them for that reason. For example, in 1961 it was estimated that plots no smaller than 1 ha, with ploughing axes of at least 200 metres and with direct connections to field tracks, were essential for mechanized ploughing to be efficient (Milhau and Montagne 1961). Such dimensions were rarely found in areas of fragmented land ownership and as the size of agricultural machines increased in subsequent years so the problem became all the more serious. Some landowners exchanged parcels voluntarily and had been doing so for decades but such operations were slow and unsatisfactory and it became clear that governements needed to pass legislation and make funds and staff available to facilitate land consolidation. The guiding principle was straightforward: dispersed plots held by individual landowners needed to be grouped into fewer, larger parcels, located as close to each other as possible. Each new parcel should have direct access to a field track to avoid enclaves being created. However, few rural planners anticipated the environmental problems that could result or the public opposition that followed particularly ruthless consolidation in several regions between the late 1950s and the mid-1970s. National approaches to land consolidation vary substantially and deserve to be considered individually.

FRANCE

In France special legislation for consolidating land in war-torn areas dates back to 1918 and a year later was followed by a more general law facilitating associations of landowners to regroup their plots voluntarily. Farms were not involved, only units of land ownership; indeed farm enlargement (for which laws were passed in the 1960s) and property consolidation remain quite separate operations in France, unlike the practice in neighbouring countries. In areas where tenancy is the custom there is a marked difference between farm holdings and property units, while in any case many functioning farms are composed of a mixture of owned and rented plots, often involving several landlords and leases spanning various lengths of time. The French version of consolidation involves removing surplus plot boundaries, laying out farm tracks and installing field drains but not building new farmsteads, as is often the case in West Germany and the Netherlands. In no way is consolidation synonymous with the radical business of land reform that has been implemented in Greece and Italy. In the late 1930s 10,000,000 ha of French farmland was judged to be in need of consolidation but only 385,000 ha had experienced this transformation. As part of its wider policy of supporting family farming, the Vichy regime passed legislation in 1941 to encourage land consolidation which remains the statutory basis for action. The law enabled landowners, farmers or staff of the local agricultural advisory service to initiate discussions on consolidation. Special committees of landowners and agricultural planners examine each proposal, which usually covers the farmland of several villages, and if they agree that consolidation is desirable then reconvene to supervise the operation. All plots that are to be regrouped are surveyed for size, soil quality and potential for mechanized working and are then awarded points so that each landowner receives his due share at the end of the scheme (Clout 1974). Public objections are received and adjustments made to the plan before consolidation goes ahead. Before 1963 up to 80 per cent of survey and reallocation costs were met by the state but all direct expenses are now covered. However, landowners have to bear a share of the cost of new farm roads, surfacing of existing tracks, improving drainage, removing hedgerows, earth banks and redundant field tracks, and clearing abandoned land prior to its return to agricultural use.

By the end of 1982 11,500,000 ha had been consolidated, namely 35 per cent of all French farmland, but a further 10,000,000 ha are judged to be in need of reallocation (sometimes for a second time) and at the current rate of activity this will take 25–30 years. By far the greatest achievements are in Northern France, where over half the farmland of seven regions has been consolidated (Ile-de-France 85 per cent; Picardie, Champagne and Alsace each 75 per cent; Franche-Comté 65 per cent; Lorraine 55 per cent). These areas include the rich innovative farmlands of the Paris Basin and its eastern fringes, with rich soils, generally low relief suitable for mechanized cultivation, and a widespread absence of hedges, walls and other field boundaries which helps keep down consolidation costs. Regrouping has produced 'neo-openfields' embracing a wide range of plot sizes, varying from approximately 1 ha in Alsace to 2–3 ha in Picardie and Haute-Normandie, and 7–10 ha in Beauce and Champagne, with local differences in the size of holdings having much to do with these results (Figure 6.1). During the 1960s and early 1970s wide stretches of consolidation were undertaken in the enclosed *bocage* lands of north-

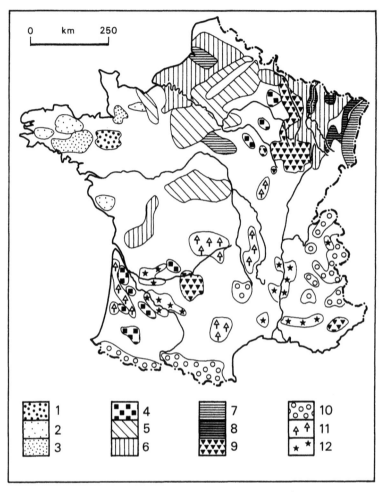

Figure 6.1 Recent changes in French rural landscapes (after Brunet 1974)

 1 Removal of vegetation from earth banks
 2 Removal of some earth banks
 3 Removal of all earth banks
 4 Important reclamation of wasteland
 5 Consolidation of openfields (plots 7–10 ha)
 6 Consolidation (plots 2–3 ha)
 7 Consolidation (plots *c.* 1 ha)
 8 Social fallow
 9 Wasteland and scrubland on calcareous plateaux
10 Marked abandonment of mountainous areas
11 Important afforestation
12 New orchards

western France, often with damaging environmental results (Brunet 1974). Much less has been accomplished in upland areas, where soils tend to be poorer, walls and hedges more abundant, and much terrain unsuitable to large farm machinery (Figure 6.2). Efforts are now being directed to districts south of the Loire but the long-recognized north/south contrast in rural France, indeed in the Community as a whole, is clearly evoked in this pattern of achievement.

The hedges and earth banks which surrounded arable plots and pastures in north-western France reached their maximum extent in the 1920s following successive waves of land reclamation during the nineteenth and earlier centuries (Flatrès 1979, Clout 1979). As well as

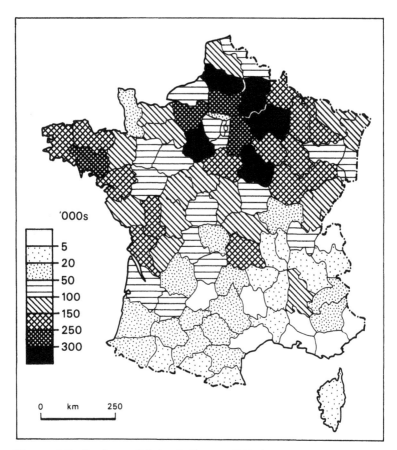

'000s

5
20
50
100
150
250
300

0   km   250

Figure 6.2   Land consolidation in France, 1980 (ha)

controlling livestock, these plot boundaries acted as windbreaks (protecting animals and fragile crops, as well as cottages), contained numerous trees (yielding timber, fruit and fodder), and were often associated with parallel drainage ditches. However they required large inputs of labour to be kept in good order and by the mid-twentieth century the inconveniences of *bocage* were becoming apparent. Mechanization was hindered; bottled gas and electricity were replacing wood as a source of fuel for cooking and domestic heating; cement was being used for construction purposes; and hedgerows afforded obstacles to drivers' visibility along country roads. Bulldozers had been introduced by the liberating US forces and enabled hedges and banks to be removed quickly and easily, while barbed wire or electric fences offered convenient stock-proof barriers. In the late 1940s plots were being exchanged voluntarily and some hedges and banks were removed but the environmental impact was slight. Official encouragement for consolidation increased during the 1950s as farmers complained about high costs of cultivating small plots in *bocage* districts, the pests and vermin that inhabited field boundaries, and the fact that up to 1 ha of farmland might be lost from cultivation along each kilometre of hedge and bank. In Finistère, with 120,000 km of hedge and bank at an average density of 180 m/ha, farming advisors investigated the impact of consolidation on local hydrology and ecology and recommended that a cautious approach be adopted (Deniel 1965).

Only small patches of *bocage* were cleared and in a piecemeal fashion during the 1950s, with considerable stretches of hedge and bank being preserved, but by the 1960s the scale of consolidation had changed and vast expanses were transformed in the name of agricultural progress. Hedges, banks, ancient trackways and ditches all disappeared as bulldozers flattened traditional landscapes and created 'neo-openfields' over much of the interior of Brittany. Agricultural planners became obsessed with straight lines as they laid out new tracks and excavated new watercourses. Farmhouses were not demolished or remodelled but gardens and orchards were cleared, along with arable fields and permanent pastures, to create 'naked' landscapes that were interrupted only by collections of debris that had been bulldozed into unsightly piles and then abandoned (to become covered with scrub) since costs had to be kept down. Construction of new highways provided a further, non-agricultural incentive for reorganizing land holdings in parts of Brittany and, of course, many other localities in the Community (Figure 6.3).

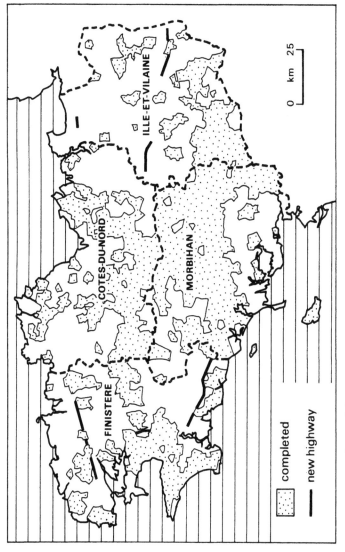

Figure 6.3 Land consolidation in Brittany, 1980

Variations in crop yield indicate strips of poor soil where earth banks used to be; however, the removal of banks and hedgerows from over 100,000 ha of the Breton peninsula has produced far more serious environmental implications (Rhun 1977). The *bocage* system played an important role in interrupting surface runoff and as soon as it had been removed localized examples of water erosion began to occur on sloping areas of farmland. In 1973 and 1974, Châteaulin, Morlaix, Quimper and several other towns experienced serious floods which swept down from surrounding hillslides from which hedges and banks had been removed. In addition, local flora and fauna became noticeably impoverished as hedgerow habitats were destroyed and artificial fertilizers and other agricultural chemicals were used on large new fields. As a result of these and other danger signals, official approaches to consolidation have been reappraised in Brittany and, indeed, throughout France since the mid-1970s (Guellec 1979). Agricultural planners have been instructed to adopt a much more 'conservative' approach and schemes implemented in the past few years have been much more moderate in dimension, with intervening buffer zones and skeletal networks of hedges, banks, tracks and old ditches being retained in an attempt to reduce runoff and environmental damage, conserve some diversity of ecological habitats, and ensure adequate provision of windbreaks (Pitte 1983). In addition, advisors are making a special effort to encourage the planting of new hedges in particularly vulnerable locations. The French institute of forestry development offers landowners a special package which comprises a metre-wide strip of black plastic which is laid across the prepared earth where each new hedge is to be created (Mills 1983). The plastic screens out weeds, enabling quite small plants to be used. Young deciduous trees are planted every 6m, with holly, chestnut and hazel being inserted between them, with species varying according to local conditions of soil and climate.

BELGIUM

Official property consolidation in Belgium followed legislation in 1956 and the first schemes bore strong similarity to the French model, involving grouping of strips in openfields and installing farm tracks across the fertile plateaux of Wallonia. Activities in low-lying Flanders were more diverse since drainage was of prime importance and so increased consolidation costs by comparison with Southern Belgium.

However, the practice of consolidation was revised substantially in 1970 and came to be modelled rather more on Dutch and West German experience to embrace the demolition of rural slums, reconstruction of worthwhile buildings, installation of piped water, electricity and other improvements. A suite of legislation followed, dealing with nature conservation and the rural environment, and culminating with a new law on property consolidation in 1977 which insisted that future schemes be undertaken with full regard for landscape, soils, topography, microclimate, flora and fauna, viewpoints and localities of particular aesthetic, scientific or historic worth (Christians 1979, 1981). Areas undergoing consolidation must be restored effectively in the physical sense, intrusive buildings and public works have to be landscaped with trees, and windbreaks installed wherever necessary. A quarter-century of activity has involved 140,000 ha (roughly one-tenth of Belgium's agricultural land), with most being achieved in districts in Wallonia with openfields and large farms, but with perhaps the most drastic environmental changes following the removal of hedges and lines of trees from the formerly wooded landscapes of Flanders and especially the Kempenland. A further 1,000,000 ha are estimated to be in need of reorganization.

THE NETHERLANDS

In the Netherlands the concept of consolidation has been interpreted even more widely to include farm enlargement and water control, since in the early years of the twentieth century nearly half of Dutch farmland was subject to waterlogging in winter (De Veer and Burrough 1978). In recent decades it has been redefined as a localized form of integrated planning which incorporates environmental matters as well as rural economies. Legislation encouraging regrouping of parcels was passed as early as 1924 but virtually all costs had to be borne by landowners. Under these relatively unattractive conditions only thirty-two projects, covering 11,150 ha, were completed by 1938 when a new law simplified administration, relieved landowners of a greater share of costs, and included road construction, land drainage and soil improvement as well as regrouping of plots (Manten 1975). The restoration of Walcheren island at the mouth of the Scheldt (which had been devastated during the Second World War) embraced wider objectives, such as construction of public housing and provision of recreation space, and conservation of special landscape features and areas of ecological value. All these points were incorporated in the 1954 consolidation act which aimed at the complete reorganization of specified stretches of countryside, even promoting local industrial schemes to employ workers who had been released from farming as Dutch agriculture modernized and mechanized. An interesting innovation was the idea of setting aside up to 5 per cent of each consolidation district for purposes of recreation and conservation. Indeed, Dutch legislators and planners became particularly conscious of the environmental impact of their work in response to widespread public outcry about the ecological implications of intensive farming systems. The rural development act of 1975 set countryside management more neatly into the context of physical planning as a whole and adopted a more open approach to decision making. The 5 per cent allocation for non-agricultural purposes remains but has been widely condemned as too meagre.

By 1960, 134,000 ha had been consolidated in the most fertile and highly fragmented districts of the country. Twenty years later the total land reached 800,000 ha and consolidation had spread into the less fertile areas of the 'high Netherlands' where landscapes included patches of woodland and heath as well as small hedged fields (Figure 6.4). The need for a sensitive approach to land management in these districts is enormous but unfortunately has not always been met. Rural landscapes throughout the country have been transformed as plot sizes have been enlarged, ancient ditches, hedges and wooded banks removed, modern land drains installed, and farm tracks widened and paved so that once-isolated farming districts have been made accessible by road and watery areas can be reached on land rather than by boat. Land sales have been strongly encouraged in areas undergoing consolidation, with the authorities purchasing individual parcels and whole properties on the free market and using them to enlarge existing holdings. Traditionally many Dutch farmhouses and barns were grouped in small villages but such an arrangement hampered the use of vehicles and expansion of buildings to meet modern requirements. The solution has been for new farmhouses and sheds for livestock and machinery to be constructed in the midst of the reorganized fields (Van Lier and Taylor 1982).

The 'area of the Great Rivers' has witnessed particularly impressive activity in order to overcome the contrast between the densely settled, highly fragmented and intensively cultivated farmland along the *levées*

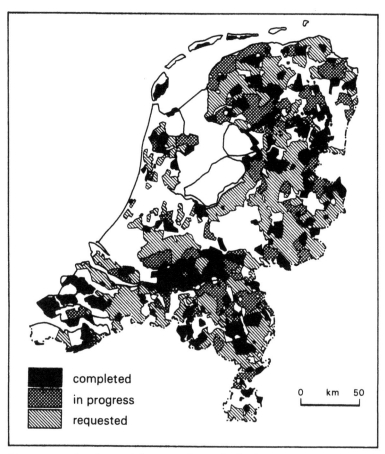

Figure 6.4 Land consolidation in the Netherlands, 1980

and the intervening basins which were too damp for settlement and were only suited to extensive pasture (Figure 6.5). Local landowners were not interested in consolidation until after mid-century but now an important range of schemes, covering several thousand hectares apiece, has reached various stages of completion (Lambert 1961, 1971). The district known as Bommelerwaard-West provides a land specimen, where farms averaged less than 8 ha and were made up of more than six widely separated plots which were only 20–30 m wide in drier areas (Meijer 1980). On average, farmers had to travel between 1 and 2 km from their farmhouses to reach their pastures and parcels

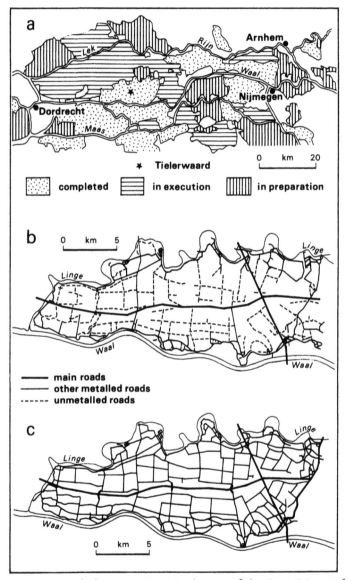

Figure 6.5 Rural planning activity in the area of the Great Rivers (after Meijer 1980)
a The area as a whole
b Road pattern in the Tielerwaard before consolidation
c Road pattern after consolidation

devoted to arable crops, fruit and vegetables. As one would expect, water management and farm roads were in desperate need of improvement. Land sales associated with the subsequent consolidation scheme have enabled farms to be enlarged substantially at Bommelerwaard-West where the number of holdings declined from 400 to 290 and many old farmhouses were demolished or converted to other uses. Over two dozen new farmhouses were built out in the newly drained basinlands and these modern buildings, constructed of bricks, concrete and tiles, stand in sharp contrast with the old houses that have been preserved in villages and along the *levées* (Figure 6.6). The total number of parcels has been reduced from 2400 to 715 and there is now an average of 2.5 parcels (each roughly 6.5 ha) to each farm. Sections of the old landscape have been conserved for fishing, walking and horse-riding, and large numbers of trees have been planted to create windbreaks and add some variety to the planned landscape. The number of cattle raised in the consolidation area has doubled, new glasshouses have been built, arable farms have been mechanized and yields have risen, while the previously empty basin

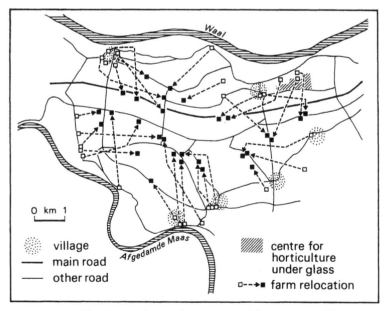

Figure 6.6 The Bommelerwaard-West consolidation area (after Meijer 1980)

lands contain their share of new houses, animal sheds and straight, tree-lined roads. From many points of view, land consolidation has brought improvement to this area and to others in the Netherlands but at the same time it has introduced monotony to the rural environment, while efficient water control, mechanization and use of agricultural chemicals have generated wide-reaching ecological consequences that many would argue are nothing short of harmful.

WEST GERMANY

At mid-century large parts of West Germany were characterized by large nucleated villages surrounded by openfields, which were fragmented into a chaos of tiny strips, with conditions being most extreme in south-western areas where gavelkind had operated. In 1952 the federal agriculture minister described a village in Bavaria with 1000 ha of farmland divided into 35,000 strips that might best be described as 'elongated pocket handkerchiefs, for even if a farmer walked along the axis of his strip when he ploughed, his cow would have each foot on his neighbours' plots' (Tracy 1982a). In villages where gavelkind prevailed, long narrow buildings were packed on to fragmented plots and some farmhouses were shared between heirs. Such conditions were quite unsuited to modern needs, indeed more than three-quarters of all farm housing in West Germany was deficient in structure, water supply or sanitation, and often with respect to all three (Mayhew 1970). Legislation was passed in 1953 that embraced everything from regrouping of plots to rehabilitation of housing and resettlement of farms beyond crowded villages (Mayhew 1973). Large integrated development schemes have been implemented in parts of North Germany but some plans have had to be simplified because of rising costs and resistance by local people to thoroughgoing change. For example, the Mooriem project on the Weser anticipated a high degree of consolidation but in fact a more modest plan had to be implemented (Mayhew 1971). Farm mechanization and improved roads meant that a lower level of regrouping was acceptable and fewer farmers actually wanted to move out of the village than had been thought at first. None the less, almost every farmer who wished to be resettled has been rehoused in or near his fields. Small clusters of new farmhouses have proved to be more acceptable economically and socially than total dispersion. Arguably the most spectacular results of consolidation have been achieved in the south and south-west of the country where tiny strips have been replaced by larger blocks of farmland and modern houses and farm buildings have been built in isolation or in small clusters along roads. As the number of village-based farms declined, so new types of resident moved in to occupy and renovate available housing.

In addition to these wide-ranging schemes, cheaper and simpler measures to exchange strips and to regroup plots but not attempt resettlement have received official backing, largely because of the magnitude of the task ahead. In the late 1970s it was estimated that almost 5,000,000 ha needed consolidation for the first time and much of the early work in Schleswig-Holstein, Niedersachsen and Nordrhein-Westfalen had proved unsatisfactory in the light of changing technical requirements and required organization. In some districts a third scheme was required and no less than 2,000,000 ha were in some kind of repeat operation (Fel: and Miège 1972). Consolidation of one type or another involved 200,000–300,000 ha each year throughout the 1960s and 1970s, when a total of 30,000 new prefabricated farmhouses were constructed outside crowded villages. In recent years rates of completion have slowed down, involving about 1000 resettlements each year and a more economical approach which involves building animal sheds outside the villages but leaving the farmhouses intact (Planck 1977). None the less, 7,500,000 ha have been subject to some kind of reorganization over the past thirty years, representing three-fifths of the total agricultural surface. Consolidation has, of course, encouraged further mechanization but in many instances the transition was far from straightforward. Many farmers had purchased small tractors and tried to use them on their tiny plots, often with little or no success; however, they soon needed to scrap them and buy larger equipment for their new consolidated fields (Schwarzweller 1971). As in neighbouring countries, field ridges, hedges, fruit trees, thickets and ancient ditches were removed in consolidation districts and criticisms of the whole process were mounting during the 1970s (Niggermann 1980). Legislation in 1976–7 responded to public opinion and stipulated that as well as assisting farm modernization, rural management schemes should be undertaken with regard to all other aspects of physical and social planning in the countryside, including local transport, flood protection, restoration of old buildings and other minor works. Improvements should be carried out in such a way as to conserve the local landscape and provide adequate space for recreation. Ecological matters have duly been taken into consideration but West German planners have made little progress in encouraging public participation in rural development (Leroy and Probst 1982).

A checklist of visual changes

In total some 20,000,000 ha have been subject to land consolidation and all that entails with respect to landscape and ecology in Belgium, France, West Germany and the Netherlands. Expressed another way, 42 per cent of their combined agricultural surface has been reorganized. Consolidation has also functioned in Greece since the late 1950s and has affected 600,000 ha, with an ultimate target being twice that amount. No less than one-fifth of all farmland in the Ten has undergone official reorganization. In other countries, farm enlargement and changes in the ownership and management of agricultural land have often conspired to produce similar results. For example, the recent tendency for traditional tenancies to disappear and for large areas of farmland to pass into the hands of financial corporations has drastically changed the agricultural activities in many parts of lowland England (Wibberley 1974). Overarching all these processes is the pricing policy of the CAP which boosts production of certain commodities and thereby encourages environmental transformation which enables increased output to take place. However, support is distributed unevenly between products and hence between regions in the Community.

Typical 'northern' commodities, such as cereals, milk, beet sugar and beef have profited from favourable prices and open-ended intervention guarantees; by contrast, fruit, vegetables, wine and other 'southern' products (with the exception of durum wheat and olive oil) have until very recently had limited guarantees and lower prices (Tracy 1982b). As a result, the productive farms of England, the Paris Basin, Belgium, Denmark and much of West Germany and the Netherlands have received generous financial support which has tended to encourage even more intensification. Less productive farmers in Mediterranean areas have experienced just the reverse and it is clear that transfers through the price guarantee section of the farm fund have contributed to a widening of the agricultural income gaps between regions in the EC. It may be argued that despite impressive irrigation schemes (often supported by EC finance) and increased output of many products, the relative economic position of Italian agriculture has not improved over the last two decades when measured against trends in many other parts of the Community (De Benedictis 1981). In addition, the fundamental dualism of Italian farming remains, with the breadth of the development gap between productive lowlands and hills and mountains (especially in the south) being little different from that at mid-century.

As a result of land consolidation and the decisions of individual farmers, agricultural landscapes in many parts of the Community have been shorn of much of their diversity and fascination, with hedgerows, small woodlands, patches of rough land and damp hollows being swept away in the name of agricultural progress (Mead 1966). For example, hedgerow removal in England and Wales rose slowly but steadily after the Second World War to reach a peak of 16,000 km each year between 1960 and 1966, and averaged 7200 km p.a. throughout the period 1945–70 (Council for the Protection of Rural England 1975). The annual rate declined to 4000 km during the 1970s but the net result has been the loss of over 200,000 km of hedgerow in thirty-five years, one-quarter of the initial total (Shoard 1980). By far the most dramatic losses have occurred in the arable eastern counties, with Norfolk losing half of its hedges between 1950 and 1970 and the figure rising to 90 per cent in Huntingdonshire, where 8000 km were removed during the same period and a 12 per cent decline was noted in just four years during the second half of the 1960s (Westmacott and Worthington 1974). However, many farmers in livestock areas have also reorganized their holdings and have ripped out extensive sections of hedgerow. The new 'openfields' are being used either for arable cultivation on a vast scale (with as few as two agricultural employees to every 100 ha) or for paddock livestock raising, with the soil being sown with special grass strains and sections of land being grazed in rotation with the help of electric fencing (Terrasson and Tendron 1981).

Isolated bushes and trees, copses and small woods have also been removed in quantity and for reasons that are broadly similar to those for hedges. Over many parts of Western Europe trees are no longer necessary for constructing buildings, for fuel and fencing, and for a host of different craft purposes in which the qualities of particular trees were appreciated: ash for wheel spokes and tool handles, beech for furniture, sycamore for domestic equipment, chestnut for posts and casks, and elm for use in damp situations (e.g. boats and coffins) (Westmacott and Worthington 1974). Nearly half of all the semi-natural and ancient woodlands of England were lost between the Second World War and the early 1970s and one-third of all small woodlands (of less than 1 ha) were cleared over the same period. The ravages of Dutch elm disease killed 20,000,000 elms (two-thirds of the total elm population) during the 1960s and 1970s and caused further massive losses, but twice as many trees were destroyed by English farmers over the same period. Intensive arable farming would seem to be linked to ash die-back, a disease that causes ash trees to lose their leaves. Surveys in England have shown that ash trees in farmland are nine times more likely to contract the symptoms than those in or close to towns and villages (Mills 1983). Free-standing bushes and fruit trees have been removed in vast numbers from the increasingly 'open' fields of Picardy, Beauce and other parts of the Paris Basin, while in Southern Europe small patches of vines, isolated olive trees and small groves have been uprooted in great number (Harvois 1978). Conservationists have rightly drawn attention to the implications of all these changes and have argued the case for adopting an ecologically responsible approach to agricultural transformation in the future (Leonard and Cobham 1977). As has been shown, agricultural authorities in several countries of the EC have responded recently by insisting that property consolidation be undertaken with due respect for the local environment. Thus, the sweeping clearance of hedgerows and trees which was under way in East Anglia during the 1960s provoked astonishment in Denmark, Schleswig-Holstein and other arable parts of North Germany where rural planners were installing new shelter belts and ensuring that basic networks of hedges and corner copses be retained to provide habitats and pathways for wildlife and produce a degree of visual variety in the landscape (Fairhall 1971).

Extensive areas of roughland have been transformed in a variety of ways. Large stretches of ancient grassland have been ploughed up on chalky soils or in damp valleys for reseeding or conversion to arable crops (Body 1982). This has certainly been the case in the damp meadows of the sandy Ems valley in Northern Germany where broad areas of pasture were drained, ploughed up and devoted to arable crops following land consolidation (Dierschke 1978). Some small patches of pasture were reallocated but they contained only a meagre variety of plants, and only 40 out of almost 100 species of grassland plants remained. The grasslands of Southern England have also come under similar attack by the plough with, for example, half the downlands of Wiltshire being ploughed up between 1937 and 1972 and one-quarter of the Dorset downs being reclaimed in the years 1957–72 (Council for the Protection of Rural England 1975). These changes are obvious enough when downland is replaced by arable fields but there have also been severe ecological implications when downland pastures have been reseeded and treated with chemicals in order to increase productivity. Application of fertilizers raises the nutrient level of chalkland soils beyond the tolerance of characteristic grasses and other plants and favours more sturdy species that duly flourish and crowd them out. In recent years many lowland areas throughout the EC have been converted to monoculture of ryegrass which has none of the ecological diversity of traditional grassland. In addition, earthworks of great archeological significance have been destroyed as ancient grasslands have been ploughed up and long-tolerated areas of public access have also been lost. Stretches of rough grassland, with occasional patches of scrubby vegetation, have been subject to similar profound changes on the calcareous plateaux of Lorraine and Burgundy on the eastern fringes of the Paris Basin. The *Société des Friches et Taillis de l'Est* was established in 1962 with the aim of reclaiming 150,000 ha out of a total of 500,000 ha for cultivation or intensive pasture (see Chapter 8). In fact, the extent of change has been rather limited because of exceptionally complicated patterns of landownership rather than technical or agronomic reasons (Brunet 1974).

Large areas of heath and moorland have also been reclaimed either for producing crops or planting with quick-growing conifers. Soon after the Second World War, when it was imperative to produce as much food as cheaply as possible, wide stretches of sandy soil in the Netherlands were treated with artificial fertilizers, deep ploughed and put under arable crops. All traces of the former system of extensive sheep grazing over heathland were removed. Very similar changes have occurred on light sandy soils elsewhere in the EC; for example, in the early nineteenth century the Dorset heaths of Southern England had formed a nearly continuous block, but by the early 1970s they had been fragmented into over 100 pieces of 25 ha or more (Council for the Protection of Rural England 1975). In absolute terms their surface decreased from 18,000 ha in 1931 to less than 6000 ha. The heaths of the Suffolk coast contracted by 75 per cent between 1920 and 1970, and 72,000 ha of heath and rough woodland have been cleared in Champagne since mid-century, thus allowing existing large farms to cultivate more land and a number of entirely new holdings to be established (Brunet 1974). Many other stretches of lowland heath have been transformed in other parts of the EC and substantial sections of upland moor have also been lost, even within the perimeters of national parks (see Chapter 7). A Mediterranean variant of this process involves the advance of mechanized cereal cultivation on to unstable hillslopes which has promoted serious rilling and soil erosion in many localities. In the past such environments had been protected by judicious planting and tending of olives and other trees but as the labour force declined so the temptation to use machinery became greater. In some parts of the Mezzogiorno the results have been little short of disastrous. Instead of following the contours, tractors have been used to plough up and down mobile slopes which has accentuated erosion. Recent attempts to cut terraces and plant trees to stop the resultant gulleying have met with limited success, sometimes being frustrated by rotational landslides (Alexander 1980).

Wetlands throughout Western Europe have been drained in great quantities in recent decades in continuation of the centuries-old practices of reclamation and land improvement. However, modern systems of canalization and pumping control water-table levels more efficiently than ever before, thereby reducing the difference between traditionally wet and dry localities and cutting down the impact of seasonal flooding. Biological communities which once flourished in the wetlands have deteriorated substantially and this trend has had serious implications for birds and other wildlife. For example, the drainage of peatlands and formerly damp, species-rich hayfields in the Netherlands has severely threatened the breeding habits of meadow birds and has destroyed food resources necessary for large numbers of migratory birds which fly over this section of the EC. Despite growing concern in ecological circles, and well-intentioned conservation arguments, the threat of reclamation still looms over many wetlands that have not acquired protected status, notably many of those adjacent to the Dutch, German and Danish sections of the Waddenzee (Matthews 1971). Research by the British Trust for Ornithology (1979) has chronicled losses in bird life in several other habitats over the last three or four decades, with mature grassland, heath, traditional farmland and deciduous woodlands sharing, to greater or lesser degree, the ecological impoverishment of the wetlands.

Further dimensions of agricultural modernization

The results of many of these changes are plain for all to see but more insidious transformations are under way as a result of modern farming practices. Use of heavy farm equipment has caused soil compaction especially in those areas of the EC with clayey and peaty soils, while deep ploughing has disturbed soil horizons (sometimes with harmful effect), and has certainly reduced micro-differentiation in pedological systems. An average of 75 kg of nitrogenous fertilizer is currently being applied on an annual basis to each hectare in the EC. Inputs in Greece (33 kg/ha) are substantially lower than the modest values for the Republic of Ireland (43 kg/ha) and Italy (59 kg/ha) which stand in contrast with remarkably high average applications in the Netherlands (240 kg/ha) (Table 6.3). Very much greater quantities of nitrogen are being used on sections of Dutch farmland, with maximum figures of 250 kg/ha being mentioned for peat soils and quantities rising to 450–500 kg/ha on some areas of clay or sand. Highest applications of phosphates and potash occur in Belgium and West Germany but use of manufactured fertilizers is undoubtedly having its most profound effect in the Netherlands. Eutrophication of soil and of ground and surface water has occurred widely and affects not only intensively worked farmlands but also surrounding districts which may still be used in more traditional ways but none the less have become exposed to irreparable damage. The ecological result has been a drastic regression of plant and animal species that depend on oligotrophic conditions, first in individual localities and then over almost the whole of the Netherlands' agricultural surface. As a result of eutrophication, over

Table 6.3   Consumption of chemical fertilizers in pure nutrient content, 1980 (kg per ha of agricultural land)

	Nitrogen	Phosphate	Potash
Belgium	128	70	114
Denmark	136	46	59
France	70	62	56
W. Germany	121	75	98
Greece	33	20	4
R. of Ireland	43	27	33
Italy	59	40	22
Luxembourg	108	51	62
Netherlands	240	41	61
UK	71	24	25
EC	75	46	44

three dozen plant species have become extinct in the country since 1950 and a further 260 are now under serious threat, by comparison with 174 at mid-century (Harms 1982). The dispersion of strong applications of agricultural chemicals and of pollution from intensive animal-rearing units is clearly the main cause.

Modern techniques of livestock farming and large scale mechanized cultivation have required new farm buildings, with government grants and loans being made available to assist their construction in all member countries of the EC. Thus, French legislation in 1966 inaugurated a great drive for constructing new animal sheds and had particular effect on medium-sized holdings run by family farmers who had previously been unwilling to enter into debt to cope with this kind of expenditure (Delamarre 1976). Similar responses, sometimes in association with property consolidation schemes, have occurred in most parts of the Community. The results have generally been positive with respect to agricultural practices but a number of criticisms have to be made. Landscape conservationists lament the intrusion of silos and other large, often prefabricated, farm buildings into the rural scene, while ecologists emphasize the environmental problems that arise from the failure of disposal systems to cope with large volumes of manure and other effluents which issue from livestock units and allied processing plant. Many such buildings are constructed without regard to established conventions affecting site and dimension which are followed by urban planners.

Renovation of old farmhouses and construction of completely new homes for those engaged in agriculture have enabled some of the shortcomings encountered at mid-century to be rectified but some observers of the country scene insist that such activities, though welcome for their role in improving living standards, have produced unwelcome visual intrusions in the farming landscape. Reconstruction in areas of the EC that were devastated during the Second World War and subsequent conflict (notably the guerilla war in Greece 1944–9) generated important crops of new housing which have been complemented by construction programmes associated with land reform in Southern Europe and consolidation schemes in West Germany and the Netherlands. Improved communications, reinforced by the spread of ideas about hygiene, have served to encourage farmers to modify millions of traditional stone and timber houses or to replace them by modern box-like dwellings built of hollow fired bricks or concrete (Wagstaff 1965). Variations in building material give useful indications of the time at which such improvements were carried out (Duboscq 1976). For instance, in the Ségalas of the Massif Central traditional farmhouse roofs were grey in colour and made of schist, but red tiles were used for improvements between 1945 and 1960 and were superseded by white 'slates' made from industrialized building materials (Meynier 1982). Many improvements to farmhouses in Greece and the Mezzogiorno have been financed by remittances from émigré workers, while part-time farmers in Western Ireland and elsewhere have devoted substantial sums to improving their homes, although sometimes the results intrude strikingly into the rural landscape and offend the purist's eye (Cawley 1983). Sympathy for careful and costly rehabilitation of traditional farm buildings is not strong in the Irish countryside and similar areas where many farming people are anxious to escape as rapidly as possible from all associations with their peasant past (Aalen 1978). Despite all these controversial changes, agricultural housing as a whole in the EC is older and less well provided with facilities than urban homes (Pratschke 1981). Plenty of scope remains for refurbishing farm dwellings and providing decent social housing, built to acceptable standards, for those engaged in agricultural employment.

Apportioning blame

In recent years the cumulative extent of all these changes to landscape, ecology and other components in the Community's countryside has started to be catalogued. While accepting that living conditions for many employed in agricultural activities need to be enhanced, it has been stressed that many aspects of rural transformation that prove to be intrusive or harmful are entirely within the letter of the law and that existing planning systems are quite powerless to control them. Writing in a particularly accusatory vein about changes that have occurred in rural England, Marion Shoard (1980) has argued that the landscape is 'under sentence of death', not from the industrialist or the property speculator but rather 'the figure traditionally viewed as the custodian of the rural scene – the farmer' (p. 9), supported by generous grants and subsidies from national and EC sources. As we have seen, similar transformations are under way elsewhere in the Community and have developed even further in the Netherlands than in England. A combination of subsidy, demand, technology and both weak physical planning and lack of common sense when it comes to agricultural matters has enabled aesthetic and ecological impoverishment to occur so widely.

But it must be stressed that not all farmers are dedicated unwaveringly to agricultural intensification; nor are all technologically advanced farmers insensitive to the vulnerability of the rural environment, as a flurry of correspondence and useful debate following the publication of Shoard's claims showed. Many English farmers demonstrated that they had retained at least some of their hedgerows, had planted small areas with trees and had taken trouble and expense to be ecologically prudent in other ways. The diversity of their reaction makes it difficult to generalize about attitudes; however, family farmers, who are personally involved in working the land and seek to achieve a reasonable profit rather than going for profit maximization, normally tend to be the most sympathetic to the theory and practice of rural conservation (Newby 1977). At the other extreme, there are 'agri-businessmen' (in the form of individuals or companies) who tend to have little time for conservation since they regard agriculture as primarily a way of making money and deem the real expertise of farmers to be in the realm of management, accounting and profit maximization rather than implementing the intricacies of traditional husbandry. 'Gentlemen farmers', who are keen to maintain a traditional lifestyle in the countryside, and 'active managerial' farmers, who are directly involved in day-to-day working of the land, tend to occupy intermediate positions along this spectrum of attitudes. In order to combat the threats that surround ordinary countryside, each member of the EC has designated various brands of cherished landscape and conservation area, but even some of these are not entirely free from assault.

7 Conserving the countryside

Varied approaches

The retreat of the countryside in the face of urbanization and the profound changes that have occurred within agricultural systems have stimulated concern for environmental conservation in each of the member states of the Community. However, that concern has been expressed in various ways and has built on differing traditions so that the ten countries now display a bewildering array of legislation and responsibilities. Objectives range from protecting rare plant and animal species in their specific habitats to conserving stretches of cherished lanscape and attempting to safeguard the rural economies that maintain them. In the first instance the task is primarily scientific and involves management policies being implemented to exclude members of the general public and allow access only to research workers or serious naturalists. Except in the most inaccessible areas high in the mountains, this kind of approach is only feasible over small stretches of terrain. By contrast, policies for the conservation of cherished landscapes often incorporate access for public enjoyment and recreation and under such circumstances management matters are far from clear cut. The simple presence of large numbers of visitors can erode or even destroy what was supposed to be conserved, while provision of required facilities will certainly modify the landscape to a greater or lesser degree. Even more delicate problems underlie these issues. For example, methods may have to be devised to persuade farmers in areas of cherished landscape to avoid transforming their working environment in ways that are occurring in non-designated expanses of countryside, while ensuring that they are not penalized financially for managing their land in a rather traditional fashion. At the same time it must be remembered that those who live in cherished countryside have the same right to expect local services and off-farm employment as people who reside in other rural areas. The simple designation of cherished areas often accentuates the attraction of visitors and may convey the impression that changes that occur elsewhere in the countryside are unimportant. This is far from the case (Warren and Goldsmith 1983).

Complexities of principle and practice surround the way that the term 'national park' is employed in Western Europe. For some countries national parks are state-owned wilderness areas where public access is strictly controlled so that flora, fauna and landscapes may evolve with a minimum of human interference. In other states they are composed of privately owned farmland where access for recreation is encouraged, although new construction and modification to existing buildings are in theory placed under some kind of control. Yet another approach involves combining both formulae, with inner areas being strictly protected but outer envelopes being equipped to cope with visitors. In short, the term 'national park' is ambiguous and its use

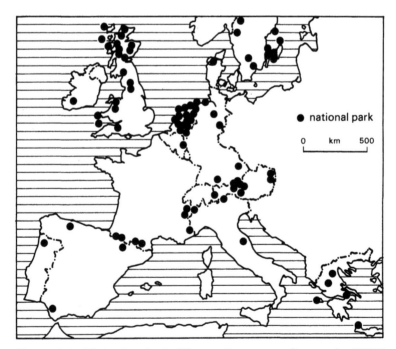

Figure 7.1 National parks in Western Europe according to the IUCN definition

conveys rather different impressions in each of the seven countries of the Community that have labelled stretches of countryside in this way. So far Belgium, Denmark and Luxembourg have not adopted this term in their strategies for rural conservation.

The International Union for the Conservation of Nature and Natural Resources (IUCN) has produced a world list of 'national parks' that makes very interesting reading with respect to Western Europe (Figure 7.1). In order to qualify for inclusion on the IUCN list, areas should be relatively large and contain one or more eco-systems that have not been altered materially by human presence or exploitation by man. They should be of special scientific, educational or recreational interest, with regard to wildlife, geomorphology or general habitat, or else possess landscapes of great beauty. In addition, national governments must have taken some kind of protective action (although not necessarily designating them as 'national parks') and public access must be restricted to those whose purposes are inspirational, educative, cultural or recreational. Such criteria automatically rule out areas officially defined as national parks in England and Wales and in the Netherlands but incorporate other rural conservation areas that do not bear that label. According to the IUCN definition, no country in the EC has more than 1.4 per cent of its total surface conserved with such a degree of rigour and in six member states the proportion is 0.4 per cent or less (Table 7.1).

Each country has its share of nature reserves that have been established at various stages in the past at the initiative of local

Table 7.1 'National parks' according to the IUCN definition

	Area of 'national parks' (km²)	Number of 'national parks'	Proportion of total area (%)
W. Germany	2,900	11	1.2
Italy	1,865	3	0.6
France	1,520	8	0.3
UK	840	19	0.4
Netherlands	590	18	1.4
Greece	280	5	0.2
R. of Ireland	43	1	0.3
Belgium	37	1	0.1
Denmark	30	1	0.1

naturalists' groups or central governments. These reserves are usually small in size and of limited access. In France the first official measure to create a reserve dates from 1853 when a group of artists argued that an area be set aside in part of the forest of Fontainebleau near Paris. Legislation dating from 1930 enabled a number of reserves to be designated within which hunting and shooting were prohibited. The Camargue, sections of the Northern Alps, coastal bird sanctuaries and inland game reserves fit into a framework of almost fifty national reserves, some of which have subsequently been incorporated in national parks or regional nature parks (Figure 7.2). Many private or local reserves have also been established. After earlier piecemeal attempts, German designation of small protected areas was encouraged

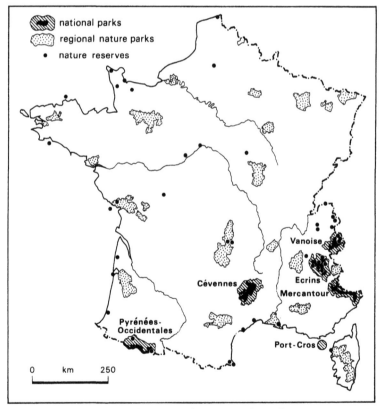

Figure 7.2 National parks, regional nature parks and nature reserves in France

by legislation in 1920 and 1935. Most of the reserves that resulted were no more than a few hectares apiece, although extensive sections of Lüneburg Heath (20,000 ha) and around the Alpine Königsee (8300 ha) were designated in 1921. The real drive for establishing large reserves with special beauty dated from the 1960s and 1970s. Now there are some 1100 nature reserves in West Germany, of which three-quarters are less than 50 ha apiece but eight surpass 10,000 ha, with such large areas incorporating sections of the Bavarian Alps, Lüneburg Heath and the coastlands of the North Sea. Unfortunately it has not proved possible to preserve a large part of the peat bogs that survived in north-west Germany until well into the twentieth century (Von Kürten 1980). National Nature Reserves have been designated since 1949 to protect scientifically important habitats in the UK. Now there are 220 of them that cover a total of 130,000 ha and range in size from 26,000 ha of the Cairngorms to 2 ha at Cothill Fen in Oxfordshire. Many are owned or leased by the Nature Conservancy Council and the remainder have been established under nature reserve agreements, whereby owners and occupiers of land agree to abide by certain conditions for use and management. In addition, 3500 localities have been designated Sites of Special Scientific Interest; forty-one County Naturalists' Trusts manage over 1300 reserves; the Royal Society for the Protection of Birds has seventy-eight safeguarded areas; and many other institutions have their own reserves.

Comparable situations may be cited in the remaining countries of north-western Europe but it must be stressed that nature reserves tend to be rather rarer in Mediterranean areas of the EC and in most instances their management leaves much to be desired. The widespread popular pastime of hunting also poses serious problems for wildlife conservation in Southern Europe, with pressures being excessively great around major cities (Raffin and Lefeuvre 1982). In France much of the Rhône valley and the north-west part of the country has reached saturation point as far as hunting is concerned but there remains scope for more shooting in the countryside of Alsace-Lorraine and in relatively empty areas in the Alps, the Southern Massif Central and Corsica (Figure 7.3). Italy has over 1,700,000 registered hunters, a long shooting season, weak legislation concerning hunting and inadequate enforcement. Large numbers of birds and other animals are shot each year and migratory birds, which fly over the Italian peninsula and islands and require safe areas to rest and feed, are particularly under threat (Cassola and Lovari 1976). For a long time Italian hunting law

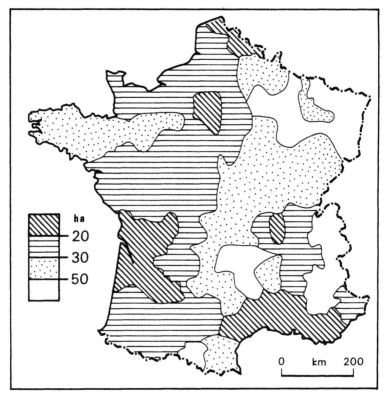

Figure 7.3 Hunting in France. Quantity of land available (ha) to each hunting permit (after Raffin and Lefeuvre 1982)

was almost entirely in favour of the hunters and against the interests of sensible conservation of wildlife (Renzoni 1975). International complaints in recent years have caused regulations to be tightened up by some regional governments, more species to be 'listed', and hunting to be banned in a greater number of localities. But protection is far from guaranteed since there are numerous inconsistencies in the legislation and practical abuses abound.

Official national parks

EARLY EXPERIENCE

The thirty national parks designated by member states in the EC vary greatly in respect of size, character and management and display a wide range of achievements and practical difficulties (Figure 7.4). The earliest and in some ways one of the more successful is the Italian Gran Paradiso park located high in the Graian Alps alongside the border with France. In the middle of the nineteenth century part of this area, well populated with chamois and ibex, was set aside as a hunting reserve which passed into royal hands in 1865. In 1919 it was donated to the Italian people and national park status was conferred on the reserve and surrounding area in 1922, although effective management dates only from 1945. Only 10 per cent is in direct state ownership. The park's 73,000 ha lie between 1000 and 4000 metres and include over fifty glaciers, with no less than 55 per cent of the area being barren and the remainder made up of high pastures, forest and scrub. Road access to the perimeter is quite good there and several villages have developed as tourist centres that offer access by muletrack and footpath into the national park (Woolmore 1971). A park association has the tasks of managing land use, providing information and camp

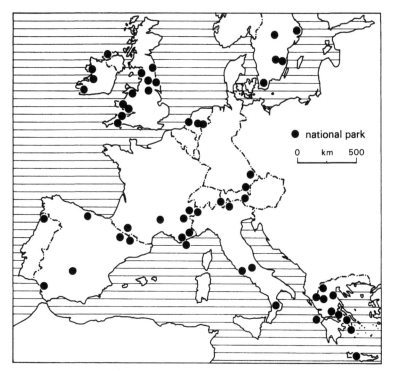

Figure 7.4 National parks in Western Europe

sites for visitors, and trying to prevent poaching of chamois and ibex. Hunting is prohibited but there is a considerable threat of poaching along the margins of the park with animals being liable to be shot as they descend to lower altitudes.

Unfortunately, conservation is much less successful in Italy's other national parks since landownership is fragmented and each experienced a phase of speculative development until management was tightened up in recent years. Just under half of the Stelvio park (95,000 ha) is state owned but the district was already very popular for skiing and other kinds of winter tourism prior to its designation in 1935. Now 10 per cent of its total area has been affected by large-scale developments for residence and tourism. Only 113 ha of the 30,000 ha Abruzzo park is state land and landownership is fragmented in Circeo park (7500 ha) which contains a town and several other residential districts (Desplanques 1973). Both of these parks have undergone a great deal of commercial development, with chalets and holiday flats being constructed both within and immediately beyond their boundaries. In fact the rate of building has slowed down noticeably since the late 1960s when management started to improve (Lovari and Cassola 1975). However, as a general rule Italian national parks have had to contend with inadequately defined boundaries, hostility from local residents (who want local roads and all other services to remain open), and absence of any kind of zoning policy, powerful speculation by land developers, insufficient support from national and local governments, and a severe shortage of trained managers and wardens. Even now public agencies are able to build roads and install power lines through designated areas without consultation with national park authorities or any investigation of impact on the landscape. Park associations need a greater voice to control development, boundaries should be recast to include areas that share an ecological unity with existing national parks, and many other shortcomings need to be rectified.

The Netherlands was the second country in the Ten to establish a national park, with the Hoge Veluwe (5450 ha) being designated in 1935 (Beynon 1979). It and its neighbour, the Veluwezoom (4720 ha) national park, occupy part of the heaths and woodlands that cover the sandy soils of the Veluwe region which contains about one-third of the country's green space. Roughly 5,000,000 people live within a radius of 100 km from the Veluwe and hence private bodies and local municipalities have had to find ways of coping with large numbers of day and short-stay visitors as well as trying to conserve the landscape. In fact up to 250,000 visitors can be absorbed in the Veluwe area on peak summer days and no less than a quarter of the Netherlands' overnight accommodation for visitors has been installed there (Anon. 1973). The Kennemerduinen park (1240 ha), between Amsterdam and the North Sea, has been managed since 1950 by the provincial waterworks authority of Noord Holland in such a way as to protect landscape, flora and fauna but also to allow visitors to enjoy the area (Roderkerk 1974). Two million people live within a 30 km catchment and some 700,000 visit Kennemerduinen each year. Management techniques have been devised to cope successfully with large numbers of visitors, with flows of 20,000 being typical for a fine summer's day (Zetter 1970). Efficiency, professionalism and adequate financial support represent the particular strengths of each of the Dutch national parks but they are, of course, far removed from the IUCN definition.

Greece was the third country to establish national parks, with Mount Olympus and Mount Parnassos being designated in 1938. Now there are ten sites bearing such a title which cover a total of 70,000 ha; however, the concept of rural conservation remains an innovation in Greece, finance is short and public interest is limited. Each park has an inner zone in which wildlife and landscape are protected by law but there is no way of enforcing regulations since only one park has full-time staff (Duffey 1982). Commercial functions, hunting and developments for tourism are permitted in the parks' surrounding 'envelopes' of land but sometimes these activities penetrate further in.

NEWCOMERS

The UK has been a relative newcomer in the establishment of national parks, with ten being designated in England and Wales between 1951 and 1957. This action followed the passage of the National Parks and Access to the Countryside Act (1949) which embraced the utopian aims of preserving and enhancing landscape qualities; protecting wildlife and places and buildings of historic, architectural and scientific interest; and providing opportunities for outdoor recreation in areas which remained the property of numerous owners and formed ordinary working space for many farmers (Sheail 1975). Indeed, established farming uses were to be effectively maintained. The ten parks covered 1,362,000 ha (9 per cent of England and Wales) and were entirely in upland areas and hence far from London and the large industrial cities of the Midlands. Planning control remained in the hands of local authorities, with a local park committee being set up for each area. These so-called national parks were administered and largely paid for locally, and management often reflected local matters rather than national priorities. They proved to be very different from what was to be specified by the IUCN, being neither in state ownership nor devoted solely to conservation. They have had to absorb a wide range of land-use demands, ranging from farming and forestry to housing, industry, mining, military training, water gathering, field sports and recreation. Three-quarters of their total surface is in private ownership and no fewer than 200,000 pepole live inside their borders.

It would be foolish to suggest that damage has not been avoided as a result of their designation but unfortunately scars have been created on the landscape of many national parks. The idea that so many funtions could be reconciled without strong direction from central government soon encountered problems (Patmore 1970). During the prosperous 1960s new demands for developing mining and providing new housing were felt and visitor pressures built up dramatically during hot summer weekends as the 'recreation boom' began to express itself. In addition, successive governments set out to raise the economic efficiency of agriculture in national parks, as elsewhere in the British countryside. Pricing policies and subsidies encouraged farmers to expand production and reduce labour inputs, which operated to the detriment of small operators and to the advantage of large farmers who took over greater stretches of upland, introduced modern techniques, reclaimed wood-land and enclosed stretches of open land inside the national parks. The number of farms within their boundaries fell by half over the past two decades, during which time the quantity of farm jobs declined by two-thirds on Dartmoor and Exmoor and by 80 per cent in Snowdonia. Lack of alternative local employment, reasonably priced housing, schools and other facilities has accentuated the decline of farm-based localities throughout the national parks, with housing being sold for holiday or retirement use and land being amalgamated into fewer but ever larger farm holdings. Many traditional farmers, with modest inputs of technology, had maintained walls and hedgerows within the parks; too many of their successors have neglected them, destroyed copses, ploughed up old meadows and reseeded them, and imple-mented many other environmental changes. In addition, commercial forestry has expanded over wide areas to provide major trans-formations of national park landscapes. For example, almost half of the 15,900 ha of moorland that had been recorded in 1933 inside the boundary of the North York Moors national park had either disappeared under the plough (2655 ha) or had been planted with trees (5075 ha) by 1971.

In general terms national parks were being exploited in exactly the same ways as other stretches of British countryside. Not only was the environment not being conserved but park administrators were failing to provide sufficient well-planned facilities for public enjoyment in a rural setting. Imaginative schemes implemented in the Peak District were exceptional but in most respects the first quarter-century of national parks in England and Wales was a period of neglected opportunity (Brotherton 1982a). Research into conditions in national parks continued and important changes in policy were introduced in 1974, following the Local Government Act of 1972. Larger finances were allocated from central government; national park committees were given greater independence and initiative; special teams of staff were set up; and a management plan was required for each park, with the specific objective of achieving its conservation (Brotherton 1982b). A decade later official policies still need to be imbued with broad conservation principles regarding technology and land use if the designation 'national park' is to be any more than a cosmetic title in England and Wales (MacEwen 1982).

Enabling legislation for creating national parks in France dates from 1960 and six were established between 1963 and 1974 (Clout 1975) (Figure 7.2). Each has an inner protected zone and an outer envelope and several contain nature reserves (Richez 1973). Four are located high in the Alps and Pyrenees, with virtually no resident population or agricultural activity in these cores, while the Cévennes park contains an established population of about 500 living on 110 farms, and Port-Cros involves a set of Mediterranean islands and adjacent water (Daudé 1976a). Entry to the nature reserves is restricted to scientists and in any case accessibility is often very difficult. Protection of flora, fauna and landscape forms the prime objective of each inner zone, which is managed by a special administrative and scientific staff in receipt of finance from the central government. Hunting is banned (although there are some exceptions) and the construction of roads, buildings and tourist facilities is forbidden. More flexible legislation has been required to allow established activities to continue in the Cévennes park which contains a markedly humanized set of landscapes, with 15 per cent of its surface being farmed, 23 per cent being viable woodland and the remainder being composed of poor timber (38 per cent) and rough grazing (24 per cent). Careful work by imaginative national park officials has enabled a working compromise to be achieved between conservation, rural tourism and upland agriculture in the Cévennes where many novice farmers accept the virtues of ecological principles (Daudé 1973).

By contrast, management of each peripheral zone is in the hands of a committee from the local *département(s)* which works toward the dual objectives of maintaining existing landscapes and ways of life, while providing facilities for visitors. These 'envelopes' of land are equipped with access roads, information points and accommodation for visitors, and these activities complement long-established functions in existing settlements where local residents are employed in agriculture, services and light industries (Aubert 1971). In general terms, conservation of inner areas has been effective, although there is mounting pressure from developers to build new roads, chair lifts and ski runs from the envelope of the Vanoise park into its core area. Already more than 250,000 visitors come to that particular cherished area each summer, where the generous provision of sporting facilities, hydroelectric equipment and associated industries in the outer envelope has arguably damaged the overall concept of the national park (Préau 1976). But all peripheral areas are subject to intense visitor pressure of one kind or another and, in any case, have to cope with the near contradictory objectives of promoting conservation and rural tourism. Management problems range from localized crowding of visitors and their cars to the erosion of pathways by hordes of hikers (Thénoz 1981). Even the islands of the Port-Cros park do not escape these problems, since 100,000 visitors arrive by boat each summer and daily totals have reached 10,000 at the peak of the season (Miège 1976).

Finally, two national parks were created in West Germany during the 1970s, involving sections of the Bavarian Alps, and others are planned, notably in the North Frisian Waddenzee (170,000 ha). The Bavarian Forest park covers 13,000 ha and contains altitudinal zones of woodland, ranging from mixed deciduous-coniferous forest to sub-alpine mountain pines on the highest points. Some low-lying enclaves have been cleared for cultivation but the whole area is owned by the State of Bavaria, as is the greater part of the Berchtesgaden national park (20,000 ha), which includes an important nature reserve and is largely wooded, although some peaks rise above the treeline. According to German federal law, national parks are primarily for nature conservation, with public access for education and recreation being permitted only so far as it is compatible with the main objective. Both existing parks are well staffed and managed, with important measures being implemented to restrict visitors to particular routes and locations, thereby encouraging plant and animal communities to survive without human intervention (Von Kürdten 1980). Reviewing the full range of national park experience in the seven countries, a clear contrast in managerial success and financial support emerges between northern and southern parts of the Community. To some extent the few national parks in England and Wales form an exception to that generalization but many would argue that they should bear another, less demanding label.

Nature parks

As well as creating national parks most countries in the EC have designated attractive rural areas as 'nature parks', which are open to visitors and often include farmland and woods with associated rights of fishing and hunting, in addition to nature reserves and protected landscapes. These parks also contain facilities for outdoor recreation, such as boating, camping and hiking, but are spacious enough to embrace habitats in which a rich and varied wildlife may survive. The idea dates back to the start of this century when the German Nature Conservation Parks Association decided to purchase stretches of countryside in order to protect them. The origins of the Lüneburg Heath nature park date from the Association's action in 1909 and 7000 ha (including 500 ha of wetland) are still managed within the park by that organization. The vegetation cover is maintained chiefly by sheep grazing and a detailed plan to cope with visitors is in operation, including large car parks and an extensive network of paths for walkers, cyclists and horse-riders. Motor traffic has been restricted to two roads and over 100 horse-drawn carriages are run for those to whom walking does not appeal (Toepfen 1981).

The popularity of Lüneburg Heath among visitors led the Association to campaign in 1956 for a suite of nature parks to be established throughout the country. The idea was felt to be utopian but it appealed to the federal and state governments and was boosted by the 1976 Federal Act for Nature Conservation. Many local authorities also found the concept attractive so that by the early 1980s more than sixty nature parks had been established, covering 5,000,000 ha or 19 per cent of the total surface of West Germany (Ruppert 1980). Not all

Länder have been equally committed to nature parks and they are mostly concentrated in the middle mountain region of the country; however, important new designations have been made in Northern and Southern Germany in recent years, while trans-frontier parks have been established in the border districts with Belgium, Luxembourg and the Netherlands (Firnberg 1971, Christians 1978). The inner sections of nature parks are used for farming, forestry and conservation, with parking places and visitor facilities being provided around their fringes (Klöpper 1976). Quite rigorous controls effectively ban any proposals for development that are deemed to be out of harmony with existing landscapes, even if such decisions conflict with local economic and social interests (Wild 1981). Not suprisingly, tensions have developed in nature parks between the authorities and private landowners, many of whom resent intrusions by the general public and are suspicious of the limitations placed on their freedom to use their own land.

The experience of Lüneberg Heath and the Dutch Veluwe was instrumental in showing how a range of land-use objectives might be implemented in specified areas. The idea of 'regional nature parks' was developing in several parts of France in the early 1960s and was formalized legally in 1967 to cope with the following aims: to provide recreation space for large cities; to support rural activities, especially in areas which do not readily adapt to modernized farming; and to protect nature and landscapes (Daudé 1976b). The decision to set up a regional nature park is not taken by the state but by local authorities (*communes*) which submit their plans for approval by the appropriate region and the state. *Départements* provide most of the finance and *communes* implement management decisions that represent delicate compromises between the three specified aims. The great advantage of this system is that decision-making is placed in the hands of the representatives of people who actually live and work in the parks, rather than being imposed from outside. But, of course, local attitudes are far from being unanimously favourable and there are examples of *communes* opting out of designated parks (Blacksell 1976).

A score of parks was set up after 1968, ranging in size from 10,300 ha at Saint-Amand Raismes, close to Valenciennes, to 281,500 ha in the Volcans d'Auvergne (Figure 7.2). Each is the product of a unique set of local circumstances and political pressures (Clout 1976b). For example, despite the designation of two nature reserves (in 1928 and 1964), the Camargue remained seriously threatened by urban expansion and industrial pollution and many local residents believed that to label the area as a regional nature park would enhance the possibility of environmental conservation (Picon 1978). Further north, the hope of supporting the local economy through rural tourism, halting firmly entrenched depopulation, preserving traditional buildings and minimizing damage to the landscape as a result of mining were key reasons for setting up the Volcans park; while conservation of a major strètch of woodland for public recreation was central to the designation of the park at Saint-Amand Raismes. Residents of rural areas close to major cities looked to the regional nature park formula as a way of organizing visitors and traffic along specific itineraries, thereby controlling widespread trampling of crops, dumping of rubbish and the threat of forest fires.

Management schemes within the parks are extremely varied, with some being equipped for fairly intensive recreational use but others being rather more similar to the outer envelopes of French national parks. For example, extensive restoration of high-altitude sheepfolds has been accomplished in the Corsican park and fire-belt pasture zones have been cleared at lower altitudes (Richez 1983). In the Alpine Vercors park young farmers are being trained to become seasonal ski-tour guides in order to enhance their income (Naudet 1976). Park authorities in Haut-Languedoc are unable to ban the rising tide of piecemeal reafforestation but they are providing strong support for modern stock rearing in an attempt to counterbalance this trend. 'Eco-museums' have been opened in several parks, along the lines of Scandinavian open-air museums, in order to display vernacular architecture, furniture, implements and vestiges of traditional rural activities. The most successful example is focused on reconstructed farmhouses, sheepfolds and pine-tar works in the Landes park (Hays 1976). Alongside these interesting initiatives the parks are encountering a number of criticisms and difficulties. Many external observers agree that there is inadequate control of land use and other physical changes within their boundaries, since new construction and other forms of development associated with established economic activities are permissible, provided local guidelines on style and building materials are respected. In addition, the financial basis of several parks is proving to be inadequate, despite the creation of a federation of regional parks in 1970 to pool certain activities and thereby keep certain expenses down.

Other measures

Beyond the designation of conventional national parks and nature parks, some countries in the Ten have experimented with other measures to promote rural conservation, land-use zoning and the safeguarding of traditional farming systems in attractive stretches of countryside. For example, Denmark has nature reserves but no national parks, yet it does possess an interesting framework of legislation and physical planning that provides strict measures to organize the use of rural resources (Newcomb 1972). Special conservation courts, which originated early in the twentieth century, enforce regulations and normal land-use planning is guided by landscape classification maps that synthesize ecological, cultural, historical and recreational information and divide the countryside into three categories according to scenic value (Poore and Ambroes 1980). Recreation facilities, summer cottages, houses and new buildings other than those necessary for agriculture are excluded from areas deemed to be 'of greatest interest' that receive special attention for promoting nature conservation. Scattered development is avoided in areas 'of great interest' unless the buildings are to be used for purposes of farming or fisheries; while construction and other environmental changes may be authorized in remaining rural areas, provided planning rules are respected. Additional legislation in 1969 included strict controls in officially recognized 'rural zones' and introduced further measures for countryside conservation (Clark 1982b). The results have been frankly controversial, producing numerous contortions in land values and often denying small settlements in 'rural zones' the kinds of modern development that they need to provide off-farm employment and retain their working population (Holmes 1973).

Paradoxically, the Netherlands possesses the most intensive and environmentally damaging farming systems in the Community, but also a particularly comprehensive framework for conservation which protects many plants and animals on both state and privately owned land. In recent years the Dutch have developed a new idea for establishing 'national landscape parks', each of which should incorporate at least 10,000 ha and comprise a combination of established farming landscapes and buildings, sites of historic and cultural importance, and nature reserves that together represent a coherent geographical whole. A provisional selection has identified almost a score of such areas which cover 200,000 ha or 8 per cent of the nation's agricultural surface. Within their boundaries it has been proposed to set up 'management areas' where farming practices would be subject to rules and regulations and farmers would qualify for financial compensation for agreeing not to use chemical fertilizers or pesticides or to adopt intensive techniques of cultivation and livestock husbandry. In addition, they would be paid for accepting to undertake a range of landscape-management tasks, which would include maintaining trees, hedgerows, copses, scrubland, ditches and pools, and also organizing their holdings in accordance with conservation principles, for example by maintaining a highish water table, preventing the digging up of meadowland, and only cutting grass after specified dates (Fornier 1979). Agriculturalists would also be required to maintain the existing external appearance of farm buildings and trackways. In fact there has been much resistance to these ideas from Dutch farmers who fear that they would have to relinquish their independence and be no more than park-keepers. Only a few have expressed willingness to enter into management arrangements of this kind. Nor are members of the Dutch conservation lobby too enthusiastic about the proposals since they believe that defending traditional uses of land in 'national landscape parks' will be accompanied by ever greater intensification of farming and further landscape deterioration in other areas, thereby isolating 'management areas' from their surrounding ecological environments.

Much more thought and effort needs to be devoted, not only in the Netherlands but throughout the Community, to finding ways of managing complete stretches of countryside in an holistic or integrated fashion and to trying to devise acceptable compromises between countryside conservation and the furtherance of economic and social objectives in rural areas. After decades of activity the total surface in the Ten that is devoted to official national parks, nature parks and other such designated areas – with their varied approaches and records of conservation – amounts to about 9,000,000 ha, with over half of the figure being made up of West Germany's nature parks. Nature reserves account for a much smaller area and contribute to a grand total of some 11,000,000 ha of officially recognized cherished land that encompasses a mere 6.5 per cent of the land surface of the Ten. The vast majority of the Community's countryside does not bear any special label and is exposed to the day-to-day activities and demands of agriculture, forestry and many other functions that operate beyond the jurisdiction of existing codes of physical planning. As such it is remarkably vulnerable (Davidson and Lloyd 1977).

8 Managing rural areas

Changing features of the CAP

Little specific recognition of the character of rural areas and the problems encountered in them has been accorded by the ten national governments or by the EC as a whole. Of course, each national or supranational decision regarding agriculture, education, transport and every other sector of life is of significance for the countryside but very rarely have these interwoven implications been articulated in an explicit way with regard to rural space. The much-criticized CAP has a very great deal to do with the evolution of the major user of land in the EC and with a declining but still politically significant employer of rural people but in no way does it amount to a coherent policy for the Community's rural areas. Before the creation of the EC each country had pursued quite independent policies in the attempt to ensure an adequate supply of food, with memories of wartime shortages remaining strong. Subsequently the Treaty of Rome specified that the future CAP should embrace a range of social and economic objectives and would give due consideration to regional disparities in agricultural activity and to variations in the social structure of farming. It would attempt to regulate prices, provide aids for production, marketing and storage, and operate mechanisms for stabilizing imports and exports, with finance being allocated through the Agricultural Guidance and Guarantee Fund.

During the early 1960s national marketing systems were harmonized, as were commodity prices later in the decade, but these mechanisms proved quite insufficient to avoid surpluses and reorientate production. Price levels were fixed high in order to support family farms but that arrangement also encouraged greater output from large commercial holdings. The guidance section of the farm fund paled virtually into insignificance by comparison with the vast sums that were spent on guaranteeing product prices, and it became obvious that fresh thinking was needed about the objectives of the CAP and especially with regard to farm structures. Agriculture Commissioner Sicco Mansholt presented an analysis of problems and recommendations for the future in his 'Plan' of 1968, whereby he advocated a sharp reduction in the Community's farm population, the replacement of small family farms by a smaller number of large modern enterprises, and a 7 per cent reduction in the farmed surface by 1980, with extensive areas being converted to forestry, national parks or recreational use. The Massif Central, Provence, Corsica and the Mezzogiorno would bear the brunt of this land-use transformation.

The Mansholt Plan and its more moderate successor of 1970 were greeted with a mixture of shock and horror by the Community's family farmers and by many national governments, since massive job creation schemes would be required if rural areas were to absorb the labour released from agriculture and, in any case, many administrations were pledged to maintain the family farm as the basic unit of agricultural production. Nevertheless, Mansholt's visions of the future heralded a range of new features which were introduced to the CAP in 1972 and which some commentators argue should have been implemented more vigorously (Thiede 1976). Following the example of France, where such schemes had been in existence for a decade, finances were made available to national authorities to implement measures for encouraging elderly farmers to retire (Directive 72/160) and for promoting the retraining of workers who chose to quit farming (Directive 72/161). The impact of the retirement scheme has proved to be modest for a variety of reasons, including administrative complications and the reluctance of farmers of all ages to give up working the land at a time of great economic uncertainty. Despite - or perhaps because of - the remarkable fragmentation of its farm structures, Italy has not participated in the scheme.

In addition, in 1972 sums started to be allocated under Directive 72/159 to assist the modernization of farms for which development plans had been approved. In order to qualify for approval farmers had to be full-time operators, have received formal education in agriculture, keep regular farm accounts, and belong to the appropriate farmers' organization in their region. Not surprisingly, a sizeable proportion of farmers in poorer agricultural areas simply did not qualify. National governments set the 'development farm' scheme in operation with varying degrees of speed. It was taken up enthusiastically in Belgium, Denmark and the Netherlands, but rather more cautiously in West Germany, the Republic of Ireland and the UK, while France and Italy took a long time to implement it (Tracy 1982a). These differences in reaction were due to many factors associated with administrative efficiency and political expediency. Only about 25,000 plans have been approved each year throughout the EC, which is a very small figure when set against the total number of farms. In the late 1970s units of 20–50 ha accounted for 45 per cent of approvals and those exceeding 50 ha made up a further 35 per cent, with many plans incorporating proposals for enlargement. None the less, most of these units were clearly family farms, since 63 per cent employed less than two persons full-time, 25 per cent between two and three, and only 12 per cent more than three persons (Calmès 1981).

In 1975 the CAP introduced a greater awareness of spatial variations through Directive 75/268 which authorized financial compensation for farmers operating in mountains, hilly terrain and other 'less-favoured areas' (Bowler 1976). The objective was to 'ensure the continuation of farming, thereby maintaining a minimum population level, or conserving the countryside' (Weinschenck and Kemper 1981). Three types of 'less-favoured areas' were recognized: those with permanent handicaps of altitude, climate, steep slopes or poor soils (e.g. Pyrenees, Massif Central, Alps, Apennines, Corsica); districts which were undergoing serious depopulation and displayed very low densities of peopling (e.g. upland Britain, Ireland, Southern Belgium, South Germany, Sardinia); and smaller areas with poor drainage, inadequate infrastructure or where it was especially desirable to maintain farming to conserve the countryside or safeguard it for specific purposes of tourism (e.g. small areas of West Germany and the Netherlands (Figure 8.1) (Christians 1977). The idea of upland farmers being 'gardeners' of rural environments which had been endowed with high social value by urban Europeans was being articulated in this third type of area, since the political influence of upland dwellers in the EC would not seem sufficient to justify the magnitude of these measures.

The directive enables governments to pay an 'annual compensatory allowance' for high operating costs in less-favoured areas. This sum is provided as a grant for each farm and is calculated according to its surface or, more usually, in relation to its number of meat livestock. Favourable rates of financial assistance are permitted for modernization of farming and also for investment in non-agricultural activities, such

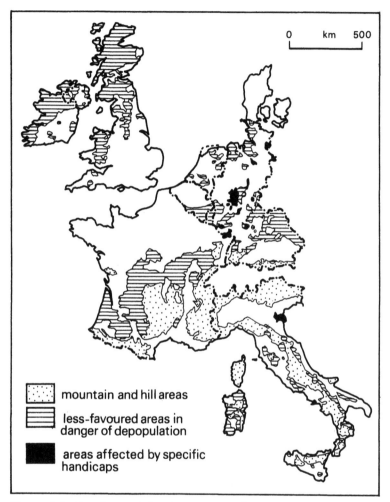

0   km   500

mountain and hill areas

less-favoured areas in
danger of depopulation

areas affected by specific
handicaps

Figure 8.1   Less-favoured farming areas according to Directive 75/268

as rural tourism and crafts, in order to diversify local employment and
reduce depopulation. A quarter of the cost of eligible measures may be
recuperated by national authorities from the guidance section of the
farm fund, and in the early 1980s one-third of guidance payments were
being allocated to less-favoured areas (Arkleton Trust 1982). Some
commentators have warned that under certain circumstances compensatory allowances, in association with social welfare payments, could

further discourage older farmers from retiring and releasing land and might thereby hinder agricultural reorganization and succession by younger men (Gillmor 1977). Others stress that many 'gardeners' of landscape in less-favoured rural areas benefit very little or even not at all from these support measures since their farms or their herds are too small (Petit 1981).

As well as allocations under the Common Regional Policy, a number of special measures have been supported from EC finance to assist less-favoured areas. These include a five-year scheme to improve livestock husbandry in Italy, a ten-year farm-improvement project in Western Ireland, a cross-border land-drainage scheme also in Ireland, and 'Integrated Development Programmes' (IDPs) for the Western Isles of Scotland, the *département* of the Lozère in the Southern Massif Central, and the province of Luxembourg in Southern Belgium. The ten-year project for Western Ireland involves improvements to land drainage, more intensive fertilization of farmland, reseeding and fencing of pastures, construction of new farm buildings and improvements to existing ones, modernizing marketing, promoting forestry, and completing electrification and provision of piped water in remote localities. The three IDPs vary in detail but share the distinction of extending beyond farming to embrace a wide range of rural activities. It is hoped that this approach will overcome the fragmented and uncoordinated nature of EC support for earlier rural projects that often reduced their effectiveness (Orlando and Antonelli 1981). The extra finance that the IDPs bring with them should enable existing schemes to be pursued more quickly and innovative ideas to be experimented with.

An analysis of trends in demography, employment and service provision has revealed the Lozère to be the most 'fragile' section of rural France (Schéma Général d'Aménagement de la France 1981). Population densities average a mere 9/km² across wide areas and the *département* contains only one town (Mende) with more than 10,000 inhabitants. Farm-based employment is three times the national average but agricultural incomes are only one-third the French mean. Off-farm jobs are rare and poorly paid, and the rural population continues to decline through outmigration and natural decrease. It is almost as if the area had selected itself for an IDP! The programme comprises three main components: deep ploughing, fertilization and general improvement of agricultural land; consolidation of holdings and construction of farm roads; and modernization of livestock husbandry. Minor features involve support for road improvements, electricity supplies, clearance of anarchic woodland to assist farm enlargement, construction of new buildings and renovation of chestnut growing in the Cévennes. The IDP for Luxembourg province in Southern Belgium builds on a range of management initiatives taken in rural Wallonia since 1978 that embrace innovations in service provision, construction of social housing and community buildings, and introduction of small-scale tourism schemes (Albarre 1982). Unlike the others, the IDP for the Western Isles has provoked particularly strong opposition from ecologists and will be discussed in some detail later in this chapter.

The balance of agricultural price guarantees remains strongly in favour of northern parts of the EC but since 1978 rather more support has been granted from the farm fund to projects in Southern Europe, including irrigation schemes in the Mezzogiorno and Corsica, farm-advisory services and co-operatives in Italy, and afforestation and improvements to rural infrastructure in the uplands of Italy and Southern France. Introduction of these schemes reflected the realization that it was necessary to readjust the balance of farm policy more favourably towards Mediterranean Europe to start to compensate for past shortcomings and to be ready to cope with problems associated with enlargement. A similar line of thinking lay behind an integrated plan for Mediterranean regions that was announced by the European Commission in April 1983. The proposal is for a six-year programme, starting in 1985, that will relate to the six southernmost regions of France, the Mezzogiorno plus areas of Central Italy, and the whole of Greece except greater Athens (Figure 8.2). Total expenditure is budgeted at 11,000,000 ECU (£6,710,000), of which 60 per cent will be contributed by all EC members and 40 per cent by the three beneficiaries. The overall aim is to raise income levels and improve job opportunities in these strongly rural areas by better co-ordination of existing Community aid through the CAP and the regional and social funds, and concentrating on specific projects. In general terms the projects involve two parts: measures to improve production, marketing and processing structures and to bring farm production more in line with market requirements; and measures to create alternative rural employment and to provide facilities to cope with weaknesses in the socio-economic fabric of these southern regions (Natali 1983). Two-fifths of the budget is to be allocated to agriculture (improving the quality of viticulture, olive growing and livestock

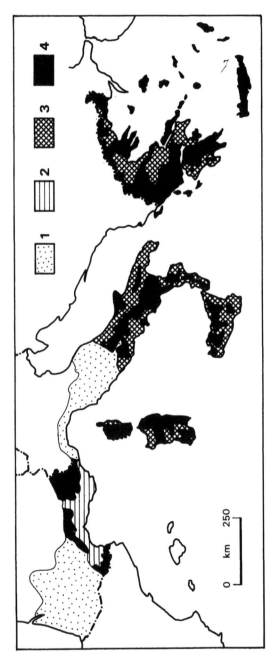

Figure 8.2   The integrated plan for Mediterranean regions

1 Inland areas (mountainous and upland zones) where all measures are applied with maximum intensity

2 Lowland areas: measures to enhance the value of agricultural products and promote conversion to new crops, fishery measures and non-agricultural job-creating measures

3 Lowland areas eligible for the same types of measure at lower intensity

4 Areas eligible for limited agricultural measures and for some non-agricultural job-creating measures

husbandry, and diversifying into fodder production); one-third to providing off-farm employment in small and medium-sized industries, craft activities and small tourism projects; and the remainder to extending forestry, modernizing fishing and supporting training schemes.

National initiatives: the example of France

As well as these measures stemming from the EC, individual countries have operated a range of management schemes in specified rural areas. For example, special assistance has been made available in the crofting counties of Scotland to rationalize landholdings and modernize farm production, and since 1965 the Highlands and Islands Development Board has functioned over a broader area and to a wider brief which includes tourism, fishing and industrial development as well as agriculture (Bryden and Houston 1976, Bryden 1981). By contrast, the attempt failed to set up a Mid-Wales Rural Development Board in the late 1960s and the Northern Pennines RDB only survived between November 1969 and March 1971. In West Germany rural programmes operate in Baden-Württemberg, in the mountains of Bavaria, and in three woodland areas (Black Forest, Odenwald and the Swabian-Franconian forest) in order to develop and support rural jobs outside agriculture. Special assistance has been made available in the westernmost parts of the Republic of Ireland, which are perceived as a link with the past and the repository of the nation's cultural heritage (Kearns 1974). Measures include special family allowances for Irish speakers, holiday stays for Irish teenagers in Gaeltacht households to enable them to enhance their linguistic abilities, incentives for tourism and small-scale industrialization to generate new job opportunities in the area, and investments in rural housing and essential amenities (Johnson 1979). None the less, numerous influences continue to increase the use of English in the Gaeltacht, including the daily impact of state-run television, with 95 per cent of its programme content in English; the fact that many government agencies conduct their business in English; the linguistic effects of industrialization (with managers and key workers operating in English and making it the language of the workplace); and the gradual influx of English-speaking residents, including many émigrés who have chosen to return to Western Ireland after periods of work in Britain or the USA.

In Southern Italy numerous schemes have been implemented by the

Cassa per il Mezzogiorno since 1950 for modernizing agriculture, providing irrigation and generating many kinds of off-farm employment. Most of these projects have been located in stretches of coastal lowland so that the mountainous interior of the mainland and the islands has been largely neglected. In the last few years, and especially since the Italian Mountain Act of 1969, a little more interest has been directed toward the uplands, which cover 4,700,000 ha and house 3,300,000 people. But in many districts rural incomes remain very low, unemployment and underemployment are rife, and in some localities the desperate search for work continues to fuel outmigration, while in others people have simply given up (Lichtenberger 1975). Local schemes are being supported by the Cassa to embrace agricultural improvement, handicrafts, small industries, camping sites and small hotels, conservation of historic sites, and improvements to local roads, hospitals and schools. The main difficulty is that the initiative for starting such things cannot be imposed from outside but has to come from local communities that have proved largely apathetic to new ideas.

Of all the nations in the EC it is the French who have been most forthright in devising plans to cope with human and environmental problems encountered in the countryside (Béteille 1981). During the late 1950s and early 1960s a number of rural planning corporations were set up to improve a range of economic and social conditions in their designated areas (Tuppen 1983). As mixed-economy companies they have been financed by public and private bodies, by individuals and the state. Major advantages include their ability to polarize local investment, build up interest in the success of the schemes and cut through fragmented administrative responsibility to co-ordinate action. Improved water management is central to the activities of six of the corporations, with irrigation and improvement of water supply figuring largely in schemes for Gascony, Languedoc and Provence-Durance, and land drainage being a vital feature of projects for the Atlantic marshes and the eastern plain of Corsica. The corporation operating in the Auvergne and Limousin includes projects which relate to agriculture, forestry and rural tourism, while land reclamation and creation of new farms have been major objectives in north-eastern France and the Landes. Between 1938 and 1952 over 400,000 ha of woodland had been burned in a series of disastrous forest fires in the latter area, where it was necessary to introduce an improved system of land management, including major firebreaks. About 150 new farms, covering 8000 ha, were established in these corridors but land reclamation, soil fertilization and water control were undertaken imperfectly and one-quarter of the newly cleared agricultural land was soon abandoned as farmers were unable to derive a living (Frelastre 1974).

As early as 1967 Brittany, the Auvergne, Limousin and a scatter of other mountain areas were identified as having distinctive problems and being in need of special assistance to reduce ingrained poverty and help reduce protracted outmigration. Together these areas covered about one-quarter of France and contained about one-third of all French farms. The precise techniques to be used for introducing rural renovation varied according to local conditions but they included special investments to promote agricultural modernization and improvements in road and telephonic communication. In addition, local plans (*Plans d'Aménagement Rural* or PARs) were drawn up for small areas of countryside by agricultural experts and representatives of local communities following legislation on physical planning in 1967 and 1970 (Le Coz 1973). Each includes an analysis of current conditions, followed by propositions for developing employment in a range of sectors, conserving landscape and organizing service provision in the immediate locality (De Farcy 1975). Most PARs cover districts undergoing depopulation and embrace the idea of diversifying and expanding the local employment base. However, some have been prepared for areas experiencing immigration and suburbanization or other forms of rapid change linked, for example, to tourism or the building of new factories, major roads, hydroelectric schemes and other infrastructures (Cros 1980).

The problems of rural decline were raised many times in French parliamentary circles during the early 1970s and the government agreed to encourage measures to try to halt the withdrawal of public services from villages and small towns (Clément 1978). Circulars were sent to all prefects in July 1974 urging them to stimulate all possible action in every *département* and enquiries were commissioned on public services in the countryside in general and on the particular problems encountered in thinly populated 'mountain areas'. That designation applied to the Pyrenees, Alps, Massif Central, Jura and Vosges, which covered one-fifth of the country and on average supported fewer than 25 persons/km², although small zones were occupied more densely. 'Mountain areas' supported 3,500,000 people (7 per cent of the total); 40 per cent of their workforce was in farming and forestry (compared with 10 per cent nationally); but their agricultural output amounted to only 7 per cent of the value of national farm production. Mountain farming was strongly orientated toward livestock husbandry, with one-third of the 10,000,000 ha of mountain land being pasture, with a similar amount under woodland. The mountains offered great potential for rural tourism but facilities were developed very unevenly, with the Northern Alps being markedly overdeveloped by comparison with other uplands.

Three main things emerged from this fact-finding about mountain areas, namely the enunciation of some general principles for rural management; the introduction of contracted arrangements for countryside planning; and the preparation of area-specific programmes (Anon. 1980b). Jean Brocard, a deputy from Haute-Savoie, was charged in 1975 with outlining measures for controlling depopulation and stimulating rural economies. Existing agricultural and silvicultural activities needed to be revitalized and schemes for co-operatives and training were strongly advocated. Abandonment of cultivated land to scrub and the growth of chaotic reafforestation were condemned and Brocard recommended that voluntary land-zoning schemes, which had been tried in the Massif Central since the early 1960s, should be extended to other mountain areas. In this way efforts could be made to define localities where agriculture might reasonably continue in the future to conserve landscapes that were of particular attraction to tourists (Brun 1978).

In addition Brocard echoed earlier proposals by Duchêne-Marcillaz in order to make the mountains more liveable. It would be essential to provide a range of new employment for women as well as men, other-wise sex-selective outmigration would continue. Serious attempts should be made to maintain a minimum presence of population in the mountains, even if that meant departing from earlier guidelines and critical thresholds for closing village schools, post offices and administrative facilities. It was advocated that special subsidies should be made available to keep vital services open, although every effort should be made to experiment in order to find appropriate ways of servicing a very small, dispersed residual population. In some localities key settlement policies might be suitable but they should only be implemented after local needs for health care, education and social and commercial services among the whole community had been investigated fully.

Official approval of measures to halt further depopulation was duly expressed and in 1975–6 special finances were made to mountain farmers according to Directive 75/268, a plan for the Massif Central was announced, and the concept of rural development contracts (*contrats de pays*) was introduced (Barthélemy and Barthez 1978). These contracts involve groups of local authorities formally agreeing to work together with help from external technical and financial services to achieve clearly specified objectives during defined periods. They must relate to areas with an element of geographical homogeneity, normally the hinterlands of small market towns or key settlements, and be preceded by serious analysis of local problems (Ferré 1981). Currently there are some 350 development contracts in receipt of government aid, involving over 8000 local authorities (*communes*) in groups of 20–30 apiece and containing a total of 5,000,000 residents.

Until the mid-1970s there had been a rapid decline in the density of schools, post offices, administrative services and public transport in rural France with, for example, nothing remaining of the narrow-gauge rail system and the length of broad-gauge track with passenger services declining from 37,000 km at mid-century to 26,000 km two decades later (De Farcy 1975). After 1976 the government instructed that these trends should be slowed; at least six months' notice should be given before any future closure and its likely implications should be investigated by a committee in the appropriate *département*. In addition, critical thresholds have been revised downwards, with a minimum number of pupils necessary for a single-class village school being reduced from 16 to 9 and even less in areas in very grave threat of depopulation. Small secondary schools in rural areas have been given a reprieve and the 400-pupil threshold is no longer automatically taken as a cue for implementing mergers and busing of pupils over long distances.

By 1979 a fairly coherent government policy regarding rural service provision had emerged and the central planning agency (DATAR – Délégation à l'Aménagement du Territoire et à l'Action Régionale) and the Ministry of Agriculture agreed to set up a special inter-ministerial organization for rural planning and development (FIDAR – Fonds Interministeriel de Développement et d'Aménagement Rural). There is some additional money but FIDAR's main role is to co-ordinate existing public assistance and adapt it to local needs and circumstances (Higgs 1983). The budget is directed toward various ends, including projects which fit into coherent, sustainable programmes, support for farmers and artisans in mountain areas, finance for rural development contracts, and schemes for providing services in thinly populated districts (Vincent 1982). Strong encouragement has been given to experimental schemes whereby, for example, postal services are combined with activities dealing with employment, social security and various other administrative matters. In the Creuse *département* of the Massif Central postmen call regularly on elderly rural residents to provide some human contact and to check on their health. The idea of using official buildings to accommodate a range of services on a part-time or occasional basis has received strong support and over 2000 examples were implemented in the French countryside between 1978 and 1980.

Local initiative has been used in order to run unconventional forms of public transport in several thinly populated areas. For example, the minibus scheme in the Lacaune mountains serves ten *communes* with a total population of 5000 distributed over 30,000 ha of the south-western Massif Central. A rota of different routes to and from the nearest market town is operated for each day of the week but customers need to telephone the driver in advance so that he can collect them from their homes and decide whether to use the twenty-seat vehicle or just a large car on that particular day (Truchis 1978). The Gondrecourt taxibus in Meuse operates a similar reservation system and conveys residents from twenty-one *communes* to the nearest town and to railway and bus stations; while the two minibuses of Treffort-Cuisiat (Ain) provide a regular service on workdays between seventeen hamlets and near-by towns and can be used for other purposes at free times (Albarre and Sottiaux 1981). These experiments in the 1970s proved successful and have been emulated elsewhere. Many initiatives for running village schools have also received favourable backing from the authorities; for example, a rather more flexible approach to defining the school year has been adopted in parts of the Alps, where heavy snowfall frequently makes pupils' journeys to school extremely hazardous. Instead of simply being told that services will be withdrawn or provided in greatly reduced fashion, country dwellers have been encouraged to suggest solutions that suit local needs. They are often given a very sympathetic hearing that may be supported by financial and practical help to implement their ideas. As a result of this change in attitudes, workable alternative solutions are being introduced in about four-fifths of the instances of closure threats to schools and other services in rural France.

Insiders and outsiders

By contrast with this intimate scale of management, involving a high degree of consultation between the authorities and local residents, there are several examples of a rather more technocratic style of planning being conceived in capital cities or even Brussels for implementation in rural regions. Reactions to this kind of approach are varied, with local residents often welcoming additional finance, although regretting the lack of consultation and sometimes being enraged by perceptions of their working environment by outsiders – whether technocrats or ecologists. While not wishing to deny rural residents a decent standard of living, conservationists are often infuriated by the way that likely environmental consequences are dismissed in countryside management plans that have been drawn up at a distance. Varying blends of these issues have come to the fore in recent schemes for the Massif Central and the Western Isles of Scotland.

THE MASSIF CENTRAL

The depopulated, poor and yet starkly beautiful uplands of Central France have been the scene of numerous rural planning measures in the past two decades, ranging from the activities of the mixed-economy corporation that started in the 1960s, through extra support allocated as a 'rural renovation zone' and then a 'mountain area', to a host of PARs and rural development contracts, to the IDP for the Lozère. In addition, in 1975 the French government announced a special plan for the Massif Central which has emerged as a rather technocratic set of forty proposals which involve all or part of seventeen *départements*, covering 7,523,500 ha and housing about 3,000,000 people. These proposals focus on the following objectives: to break down the region's isolation (by financing new roads and modernizing railway links, local airports and the telephone service); to encourage small industries and tourism in rural locations; to modernize forestry, craft activities and farming (through co-operatives and better marketing systems); and to enhance services available in small and medium-sized towns (Bouet and Fel 1983).

Improvements in communications head the list of achievements to date and include a new motorway between Saint-Etienne and Clermont-Ferrand, with a second motorway to be completed northwards from that city by the mid-1980s (Figure 8.3). Main roads around Clermont-Ferrand and Limoges have been upgraded and improvements have also been made on east/west roads. Journey times to Paris have been shortened following electrification of the railway from

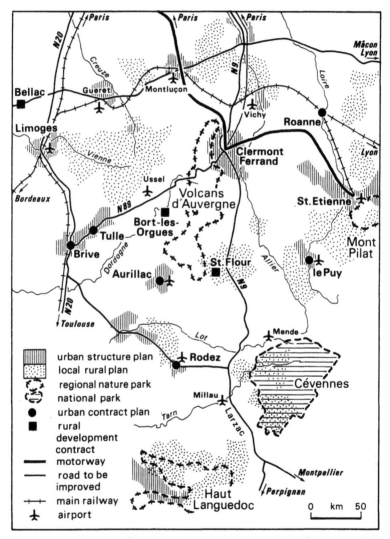

Figure 8.3 Aspects of regional management in the Massif Central (after Bouet and Fel 1983)

Clermont-Ferrand and internal air services have been improved. More than 500 firms have received subsidies to assist the creation of 17,000 new jobs in the Massif and a special industrial development association now operates to stimulate new employment; however, agriculture and especially old-established industries continue to shed labour alarmingly. Craft industries and rural workshops also qualify for special support and the Socialist government allocates the highest rates of employment premium to the Massif and other mountain areas.

Finance has also been directed to increasing accommodation and other facilities in the countryside and for modernizing fifteen spa resorts in the Massif. Other investments involve schemes for improving the quality of everyday life in the countryside, such as construction of additional social housing in selected settlements, building relay transmitters to serve areas where television reception is poor, provision of a better telephone service, and extra snow-clearing teams and equipment for districts that are cut off in the depth of winter. As a result of the 'Plan', the Massif has acquired the status of being an assisted area to a greater extent than ever before. Just how successful these predominantly technological measures will prove to be in tackling the essentially human problems of the region remains to be seen. Certainly it is clear that the large investments are concentrated in its northern section, with much more modest, dispersed schemes receiving support in the more remote and impoverished areas further south. The most recent census results indicate that the Massif contained 800 more people in 1982 than in 1975 but seven of the eleven *départements* entirely within its limits continued to undergo decline. Nine registered more deaths than births and most rural areas continued to experience net outmigration, with growth being restricted to a limited number of towns.

One small part of the Massif has witnessed one of the most sharp confrontations between central authorities and local residents – with the backing of ecologists – over the use of rural land. The affair began in 1971 when the military authorities announced their intention to extend tank firing ranges from 3000 ha to 17,000 ha and perhaps even 30,000 ha on the sparsely populated Larzac plateau near the town of Millau in Southern Aveyron (Pichol 1978). About 100 sheep farmers were to be compensated for losing their property and some 500 people and perhaps 15,000 ewes would have to be evacuated. Instead of this being simply carried through as a matter of course the local farmers protested and their case was taken up by Occitan regionalists and ecologists and anti-military groups throughout the country. Committees for the defence of Larzac were created in many parts of France and money was raised to purchase empty farms which the military had seen as easy targets for requisition.

Here was a classic contrast in perception, with the authorities in Paris appraising Larzac as dead, empty land without interest or value. The ecologists knew otherwise and demonstrated that depopulation had enabled farm enlargement to take place so that sheep rearing was in the hands of young farmers who had joined together in local co-operatives and were introducing modern techniques of livestock raising and milk production for Roquefort cheese (Pilleboue 1972). Small flocks, involving about fifty animals apiece, were on the decline and were being replaced by larger flocks averaging 150 ewes. New animal sheds had been built and small areas had recently been brought into cultivation, worked with modern machinery and used for growing cereals. Farmhouses had been improved and surplus property was being used as accommodation for holiday visitors or as second homes, especially by Belgians, and this trend brought life and additional income to the area in the summer months (Balsan 1973). To extend the army ranges would destroy all these achievements. Demonstrations against the scheme were organized on Larzac and in many other areas during 1972. The farms were thrown open to interested visitors at Eastertime and Larzac sheep were taken to graze near the Eiffel Tower when the public enquiry started in October (Ferras, Picheral and Vielzeuf 1979). Next summer 50,000 people attended a great meeting on Larzac and the dispute became a kind of national crusade, with thousands of sympathizers visiting the area each summer to demonstrate their solidarity and be lectured on ecology and the evils of militarism (Ardagh 1982). Neither side abandoned its position until the Socialist administration came to power in 1981 and promptly announced the cancellation of this widely unpopular proposal.

THE WESTERN ISLES

The current debate concerning the Western Isles IDP displays a different arrangement of interests, in which ecologists and most local residents are not in accord. The 200 km line of large and small islands in the Outer Hebrides arguably forms the most beautiful, and certainly the most remote and least urbanized area of Great Britain, where a sense of local identity remains strong. The islands cover 30,000 ha and house 30,700 people, about 6000 of whom reside in Stornoway (Adams 1983). The remainder live in small crofting townships mainly along the Atlantic coastal fringe where fertile land is found in the form of machair pastures, which contrast with stretches of blanket bog that cover higher areas (Caird 1975). The machair is sited between the hills and stretches of calcareous sands and fragmented shells along the coast. It is the product of human activity, being fertilized by grazing livestock and by seaweed carried from the shore, with its ecology depending on the level of the freshwater table. Although nominally a grassland, the ankle-deep vegetation of the machair in summer is almost pure flowers in the absence of chemical fertilizers. As well as forming the living and working environment of the crofters, the machair supports numerous species of breeding birds, many of which have been listed in the EC Directive on the Conservation of Wild Birds. For example, the tiny area of the Hebrides holds two-fifths of Britain's corncrakes, one-eighth of all ringed plovers and high proportions of many other birds. The machair, peatlands, moors and lochs of the Western Isles together support a rich selection of rare breeding birds, red deer and otters, representing a fauna 'unsurpassed in Great Britain' (Pickup and Fairclough 1982, 24).

However, this unique and ecologically rich environment is also one of the poorest in financial terms. Despite a succession of subsidies and forms of assistance the Outer Hebrides have experienced population decline, after reaching a peak half a century ago, and those who remain have a low standard of living in material terms (Bradley 1982). Underemployment, poor housing and a high proportion of elderly people represent the main social characteristics of the islands (Edwards 1983). Nearly one-quarter of all dwellings are officially classified as sub-tolerable, 21 per cent of the population are pensioners, over one-fifth of whom live alone, and there is double the national average of people aged over 75 years. Not surprisingly, reliance on social transfers is strong. There are 6000 crofts, each with a patch of in-bye land and a share of common grazing, but most are small and half are less than 5 ha in size. About three-quarters of the total area of the Western Isles is under crofting tenure (Abercrombie 1981). Raising cattle and sheep accounts for most agricultural income but animals have to be sold before the winter sets in since fodder is short and imported feed is costly. Conditions are such that crofting has become a variety of part-time farming, with 85 per cent of all crofts providing less than two full days of work each week. Tweed-making and catering for visitors provide some support, but farming and fishing are declining activities and attempts to introduce light industries and attract entrepreneurs into advance factories have been largely unsuccessful. Rates of male unemployment run at 30 per cent and are rising, and almost half the available jobs are in professional and administrative services (Table 8.1). Wages are roughly one-fifth lower than on the mainland and there are few prospects for school leavers, hence many young people move away.

In April 1981 the European Council of Ministers agreed that the Western Isles should receive finance for an IDP, which became operational at the end of 1982 with the key objective of raising the social and economic well-being of the local population. Some £20 million are to be invested for modernizing farming and fishing over a period of five years (Table 8.2), with an additional £36 million for spending on roads, piers, ferries, rural tourism, energy, crafts and light industries. The European Commission is committed to covering 40 per cent of agricultural costs from the farm fund, with the remainder being met from British sources, while non-agricultural projects are supported by the Community's regional fund. No monies are allocated for

Table 8.1 Employment in the Western Isles, 1982 (per cent)

Professional and administrative services	46
Textiles	13
Agriculture and fisheries	11
Construction	9
Distributive trades	9
Transport and communications	6
Others	6

Table 8.2 IDP investment in agriculture and fisheries in the Western Isles, 1982–6 (£ million)

Fishing and fish farming	9.0
Land development	3.4
Livestock development	3.4
Infrastructure for agriculture and fisheries	1.9
Education, training and advisory services	1.5
Comprehensive agricultural development	0.5
Costs of project team	0.3

environmental protection. Many of the features of the IDP, and especially the offer of so much 'Euromoney', have been welcomed for one reason or another in the Outer Hebrides but ecologists have profound fears regarding the implications of the agricultural investments which aim at creating efficient and viable farms.

Possible developments will involve arable cultivation, silage production and grassland improvement, regeneration of moorland grazing, land drainage, and further amalgamation of crofts. Arable production might be extended into derelict croftlands but abandoned machair soils are delicate and respond best to seaweed and other traditional fertilizers. Use of artificial fertilizers and herbicides would not only be detrimental to flora and fauna but would ultimately impoverish soil quality. Increased production of silage would involve replacing sections of ploughed machair and ecologically diverse wet meadows by areas that had been drained, treated with chemicals and reseeded with just a few grass species. In fact, reseeding of local moorlands has been quite extensive in recent years and involved applications of fertilizers and a limited range of grasses, inevitably dominated by perennial rye grass and white clover. Land drainage in the machair is not to be undertaken by piped underdrains but by improved ditches which will lower the water level. If undertaken carefully, this procedure can enhance prospects for cultivation but runs the risk of depleting productivity in times of summer drought and, of course, has profound implications for flora and fauna. Amalgamation of crofts and greater co-operation between working units will erode the existing mosaic of habitats.

The IDP will cover up to 85 per cent of the cost of such land 'improvements' in the future which have admittedly operated over tiny areas since the 1950s (Caird 1972). However, there have been profound criticisms of such activities on the island of Mull where there is a thinned-out social fabric with far fewer people living on enlarged holdings. The IDP will accelerate environmental changes on rearranged crofts and runs the risk of destroying unique habitats, with one-fifth of the machair and half the in-bye land (about 13,000 ha) likely to be drained and reseeded. By summer 1983 there had been 2000 applications for land-improvement grants, especially for fencing, and others had been submitted for schemes to improve cattle raising. The Nature Conservancy Council had scrutinized a number of applications and complained about the implications of very few of these but ecologists continue to voice their fears and their anger that the well-intentioned IDP could be implemented with so little regard to the full spectrum of environmental matters.

In the bitter words of Rose (1982), the IDP is 'an ill-conceived poisonous plan' which, he insisted, is likely to destroy the wet machair and the long-term prospects of craft industries and tourism, thereby increasing unemployment even more and disrupting stable crofting holdings (p. 20). He continued: 'We do not doubt that the Isles need money; but the IDP is a blunder not a boon' (ibid.). Real solutions to the problems of the Western Isles, it could be argued, should create economic wealth through conserving ecological wealth but there is no clear indication that the IDP will assist either objective. A radical ecological viewpoint may be too extreme for many people but few could surely disagree that under any circumstances intricate examination needs to be made of local conditions and hard thought directed to the full range of economic, social and environmental implications that introducing new technology at the behest of outsiders or even progressive farmers is likely to have for the long-term future of the countryside. If this kind of scrutiny were undertaken it is not impossible that a rather less intrusive brand of rural management might be deemed appropriate in many instances.

9 Toward a rural policy

Typology

The evidence of preceding chapters has shown that facts of geography and history and trends in society and economy make the rural areas of the member states of the EC remarkably diverse, not only region by region but also at a much closer scale, where micro-features of location and degrees of local initiative make themselves felt. As a result, each stretch of countryside is unique in its visual appearance, social composition, combination of activities and inherent problems (Rainelli and Bonnieux 1978, Fruit 1983). While fully accepting the importance of this uniqueness, it is possible to move from individual parameters (such as farm size, population structure, land use, economic activity or income) in order to identify a number of generalized types of countryside and to characterize the kinds of human and environmental problems that are encountered within them (Calmès 1978). Classification may be achieved in various ways; however a three-fold distinction between 'dynamic' countrysides, those that occupy a median position on the scale, and others that are 'in the process of abandonment' offers a reasonable overview (Livet 1965, Gervais, Servolin and Weil 1965).

Dynamic rural areas display high degrees of economic and social well-being and embrace three kinds of situation. The first involves areas of truly commercialized and highly remunerative farming, where holdings are highly responsive to market forces, have large inputs of capital rather than labour, are likely to be highly mechanized, and achieve high productivity by intensive application of artificial fertilizers and agricultural chemicals. A second situation embraces peri-urban stretches of countryside which contain homes for daily commuters and whose settlements and landscapes function increasingly as parts of the 'dispersed city'. A third variant on the dynamic theme is represented by rural areas which are highly responsive to catering for day visitors and tourists.

Median areas display more modest levels of economic and social development and support rather less productive forms of agriculture but none the less have certain potentialities for increased dynamism in the future. They are likely to be located around the fringes of generally dynamic areas or exist as enclaves within them.

Rural areas in the process of abandonment reveal low densities of peopling and have experienced long histories of depopulation that have taken generations of young, potentially innovative people to urban destinations (Boutet 1973). Their physio-environmental resources of soil, slopes and climate vary greatly but tend to be unsuitable for intensive agricultural production. Wide stretches are likely to have been afforested or simply abandoned. Such areas have the distinction of containing the least humanized and often cherished landscapes of the EC.

Despite their inherent differences, human and environmental problems are encountered in each of the three types of area. Average conditions in dynamic areas are generally healthy and incomes are high, as is the value of land and the productivity of farming. However, average indicators are misleading and one must not overlook the presence of sizeable minorities, including the poorly paid, the elderly, the car-less and other less privileged residents who may well be in need of local services and facilities which the affluent majority will simply drive to the nearest town or city in order to purchase for themselves. Pressures on land for the construction of new homes will be particularly great in dynamic countrysides and can lead to piecemeal urbanization unless there is effective physical planning. Numerous other tensions exist over resource allocation; for example there may be serious competition for housing between local residents (especially young couples) and incomers, be they daily commuters, retired migrants or owners of weekend cottages. Ecological changes are likely to be intense, with high-technology farming requiring intensive use of chemicals, the removal of hedgerows and other micro-features, and the construction of buildings for factory farming. But these very areas are also the most accessible by city dwellers, whose recreational interests are not well served by industrial agriculture (Fitton 1976).

Competition for land is weaker in median areas and ecological modifications associated with agriculture are less pronounced, but average incomes tend to be lower than in dynamic countrysides.

However, it is the thinly populated areas in the process of abandonment that display the most acute economic and social problems (Roux 1977). They contain an above-average proportion of elderly local people; despite subsidies, incomes from farming are low; there is often little alternative employment; and services of all kinds are particularly vulnerable to closure (Shaw 1979, McLaughlin 1981). These are all symptoms of impoverishment, as is the haphazard abandonment of farmland to scrub. Installation of new industries and development of rural tourism, retirement migration, second homes and 'back to nature' farming provide examples of innovation, but these are often surrounded by controversy and may generate new problems in their wake. They reflect a complex revaluation of space by some members of urban society who are fascinated by remoter places and the attractions of exploring repositories of nature and traditional culture (Gold and Burgess 1982). Designation of national parks and similar areas is in part an official expression of this kind of popular trend. However, such countrysides are ecologically fragile and large influxes of visitors and provision of facilities for them can easily damage cherished landscapes and modify – or even destroy – what little vibrant rural life that may have survived in these depopulated localities.

As well as variations between rural areas and the problems encountered in them, there is clear evidence that 'peripheral' (and that means predominantly rural) regions have suffered particularly harshly since the mid-1970s, with the gap between 'rich' and 'poor' regions in the Community having widened rather than narrowing (Cardol and Van Engelenburg 1980). Over the same period, the standard of living of progressive farmers has risen but inequalities within the agricultural sector have continued to broaden to the detriment of producers in depopulating uplands and other marginal areas (EC 1980). None the less, the human and environmental characteristics and problems of the countryside have rarely been envisaged by administrators and policy-makers in the member states in a wide, integrated, truly ecological way which transcends the confines of compartmentalized responsi-bilities. Nor has that challenge yet been faced at the scale of the EC. Of course each common policy has implications for the Community's countryside, but those relating to regional development, social matters, the rather nebulous concept of 'the environment', and agriculture are the most immediately relevant. Unfortunately the first three of these are small and relatively impecunious affairs which are dwarfed by the massive and much-criticized CAP. For example, the whole Community budget for 1982 amounted to 23,260 million ECU, of which appropriations for the guarantee section of the Agricultural Guidance and Guarantee Fund were 13,143 million; credits for the guidance section 765 million; regional intervention (including the European Regional Development Fund) 2000 million; and social measures (including the Social Fund) 1593 million (Tracy 1982b). Given this massive emphasis on guaranteeing farm prices it is not surprising that the role of other Community mechanisms for promoting employment in 'problem areas' and establishing guidelines for improving the setting and quality of life remains modest.

Most finance allocated by the ERDF (established in 1975) goes to 'infrastructure' (including roads, railways, ports, airports, water supplies) and to industry and services. Some of these infrastructural developments assist rural areas, although it would be hard to isolate the benefits that they derive. The most directly relevant grants are those that relate to infrastructures itemized in Directive 75/268 on less-favoured areas (e.g. rural electrification, water supply, access roads to farms) provided that the less-favoured area in question is also designated for regional aid by the appropriate member state. Schemes have been assisted under the 'quota' and 'quota-free' sections of the ERDF, with the latter being of particular interest to ideas for exploiting the indigenous potential of regions through emphasis on small and medium firms, handicraft enterprises and small-scale tourism that are all emphatically present in rural areas.

The main activity of the Social Fund relating directly to rural areas is its programme to assist retraining workers leaving farm employment. Of recent years, Italy has been the largest beneficiary, followed by France and the Republic of Ireland. As far as the countryside is concerned, the Community's environment policy relates to land-use change, wildlife, the implications of factory farming, use of pesticides and mineral fertilizers, and the desirability of taking ecological matters into account in regional management (Ellington and Burke 1981). Of particular interest – and great significance – is the elaboration of environmental impact assessments as a prerequisite to authorization of large development schemes (Wood 1982).

The expensive CAP embraces economic and social objectives in relation to commodity prices and farm structures (with a clear weighting on the former) but has not been required to encompass the numerous rural matters that surround agricultural production. Only in recent years has a gentle start been made on establishing area-based schemes that recognize interrelationships between countryside problems. These include the 'compensatory allowance' for farmers in mountainous and other less-favoured areas (Directive 75/268), the emerging 'Mediterranean package' of measures announced from 1978 onwards, and the 'Integrated Development Programmes'. In short, the Community has an extensive and complicated array of mechanisms, of greater or lesser relevance to the alleviation of problems in rural areas. But these instruments have come into existence at different times and for varying reasons, and their effectiveness in terms of rural development is uneven. In their present form they most definitely do not represent a co-ordinated policy for rural areas in the Community (Tracy 1982b).

Critique

Certainly members of national assemblies and of the European Parliament who espouse a wide range of political persuasions have not been reticent in criticizing the present CAP and suggesting ways whereby it might be improved. Of special relevance is the widespread concern that – regardless of its size in the future (and many would argue that it should be trimmed down substantially) – the CAP should also tackle related problems and thereby become an agricultural *and* rural policy. Numerous factors play a role in explaining this kind of advocacy. Despite fundamental changes since the buoyant 1960s, large sections of the Community's countryside continue to be converted to urban uses. Counterurbanization, on both a permanent and a short-term basis, exerts pressures for new housing and facilities in many rural areas, while ecologically fragile environments are exposed to devastation as a result of intensive farming, tourism and many other forces. Rates of urban unemployment continue to mount and rural-urban migration has dwindled to a pale shadow of what it was in the 1950s and 1960s. Two institutional matters draw further attention to problems of the countryside: first, the Community is required to draw up a new agricultural policy for the remainder of the 1980s; while second, the future entry of Portugal and Spain will produce profound changes in the composition and character of the EC (Tracy 1982b). Enlargement from the Ten to the Twelve will increase the Community's total area by 59,600,000 ha (+ 36 per cent), its farmland by 35 per cent, arable surface by 38 per cent, woodland by 55 per cent and area devoted to permanent crops (vines and olives) by no less than 92 per cent! Spain alone has 2,800,000 ha of irrigated land and entry will increase the Community's total by 80 per cent (Bethemont 1977). In addition, there are plans to irrigate a further 300,000 ha in Spain in the immediate future. Portugal and Spain together house 47,314,000 people – raising the EC's total by 18 per cent – and contain 3,181,000 farmers and labourers who will inflate its agricultural workforce by 37 per cent (Tsoukalis 1981). Enlargement will augment the volume of cereal production in the Community by one-eighth but typical 'Mediterranean' crops will contribute much more drastic increases (wine, fresh vegetables and tomatoes each up by one-third, and citrus fruit by 90 per cent). Each common policy and especially those relating to agriculture and regional development will have to be reformulated in order to take the enormous rural implications of Iberian entry to the EC into account (Tovias 1979).

The thread of argument on rural policy in France has been examined in Chapter 8, while in the UK a House of Lords Select Committee (1980) argued that the existing CAP would be quite unable to meet the challenge of Iberian entry nor would it permit the potential of rural areas elsewhere in the EC to be realized. Along with other experts, members of the Select Committee insisted that it would be unrealistic to propose any kind of rural policy that would operate in a uniform way across all stretches of countryside throughout the Community. Rather they advocated a sophisticated and spatially selective approach and recommended that a share of the finance that had traditionally been directed exclusively to farming should be devoted to stimulating and supporting off-farm activities in rural areas. Desirable objectives should include responsible exploitation of resources, promotion of employment and social well-being in the countryside, conservation of landscape, and provision of access and facilities for compatible forms of rural recreation. The Select Committee further recommended that additional IDPs should be introduced and a special rural development fund be established to support projects in member states. At a national scale, in 1982 the Agricultural Committee of the UK House of Commons went so far as to propose the creation of a new post of Minister of State for Rural Affairs who would not necessarily be attached to the agricultural ministry, which the Committee believed to lack the scope to supervise a more widely based rural policy (Higgs 1983).

In the chambers of the European Parliament Edgard Pisani presented a lengthy critique of the CAP to the agricultural committee in

February 1980 (EC 1980). This included a thorough review of social and economic problems in rural areas and advocated that 'structural agreements' should be drawn up between regions of the member states and the EC. They would extend beyond agriculture and cover rural activities as a whole, including infrastructure, education and training, and encouragement for small and medium-sized industrial and commercial undertakings, rural tourism and forestry (Orlando and Antonelli 1981). The recommendation was clearly inspired by French experience since the mid-1970s. In May 1981 Sir Henry Plumb presented a resolution to the Parliament in which he recommended combining agricultural structural projects with schemes that would be financed by the regional and social funds to create off-farm employment in rural areas and restrain further depopulation (EC 1981a). In November Edgar Faure reported at length to the committee on regional policy and planning in the Parliament on the contribution of rural development to overall territorial management and advocated that an integrated approach be adopted to countryside matters embracing farming, conservation, rural settlements, employment and local services (EC 1981b). In January 1982 Madame Barbarella, on behalf of the agriculture committee, responded to the Faure report and, while expressing some criticism, extended warm support to the idea of integrated development schemes in the EC's most backward rural areas (EC 1982). One month later the European Parliament adopted a resolution to assist rural development, with the parliamentary assembly of the Council of Europe having done just that in the final days of January.

Recommendations

In the spirit of these supranational opinions and many recommendations from organizations in individual member states it is clear that there is a strong body of feeling that in the future the countryside of the Community should be managed in a non-compartmentalized way. For example, an international group of rural managers recommended that in the light of changes under way in rural Europe, it is now time for governments to draw up rural development policies, and possibly even establish separate departments for rural affairs (Higgs 1983). Likewise, they argued that a coherent rural development policy is also needed at Community level, with much closer contact being desirable between the EC's agricultural, regional and social funds. The case for a new approach to countryside management has been expressed clearly in the UK where no fewer than eight rural organizations, as diverse as the National Farmers' Union, the National Union of Agricultural and Allied Workers, and the Council for the Protection of Rural England, have come together, have managed to play down their very real differences of interest, and have spoken with a single 'rural voice' in recent years which calls to save the vitality of rural areas and demands a positive and integrated response from government to achieve that end. However, they, and comparable groupings of rural interests in other member countries of the EC, are not suggesting that a single set of measures should be applied uniformly, since it is fully appreciated that regional conditions vary greatly and that the established rights and responsibilities of national governments and local authorities must be respected. However, it is appropriate to advance a series of broad guidelines which, after consideration by regional, national and supranational policy-makers and due consultation with local elected representatives, might be tailored in accordance with particular rural circumstances. These recommendations would be ecological in scope, recognizing that rural phenomena are interrelated in space and that rural resources are vulnerable to devastation by thoughtless development at an inappropriate scale or in an inappropriate place (Sachs 1978, 1980). They reflect many of the worthwhile, innovative schemes that are already in existence in specific areas or across individual member states of the Community. The future of agriculture, off-farm employment, and provision of rural services represent major topics of concern.

AGRICULTURE

Modernized production and financial support have improved incomes and living standards for many members of the diminished agricultural workforce. Technological developments continue to offer important possibilities for increasing the productivity of many who work the land; however, all too often the social and ecological costs of mechanization and intensification have neither been fully recognized nor taken into account (Peper 1969). Rather than continue to place exclusive emphasis on the potential of intensification it would be prudent to explore the opportunities offered by a variety of types of agricultural technology (Commins 1980). As well as firmly established forms of modern production there is surely also room in the EC of the future for types of production which pay greater respect to ecological principles, are rather more labour intensive, conserve rather than destroy farming landscapes, and place less stress on capital investment, use of agro-chemicals, maximization of output and industrialization of production (Buttel 1980, Jansen 1974).

While respecting the wishes of rural families and individuals regarding whether they prefer to continue country living or migrate to town, it may be deemed prudent to retain more people in medium-technology farming in the future rather than encourage them to join the growing ranks of the urban unemployed through continued implementation of existing versions of agricultural and regional policy. At the time of writing 11.5 per cent of the Community's labour force is out of work, with 14 per cent being exceeded in Belgium, the Netherlands and Ireland, and two other countries registering above-average conditions. Joblessness far exceeds average values in many localities – urban and rural alike – and there is little chance of reducing unemployment rates in the future. Indeed they may well rise substantially and our long-held concepts of work and non-work may well have to be revised profoundly. As well as its hosts of unemployed the EC has many thinly populated rural areas and future policies relating to spatial disparities should pay attention to *both* of those facts.

New emphasis should be placed on finding ways of keeping young, potentially innovative people in rural areas and enabling a proportion of them to enter farming so that custody of the Community's agricultural resources may be in capable hands in the future.

Multiple job holding, combining on- and off-farm employment, might be viewed with greater favour as a way of supporting residents in rural areas and bolstering demand for local services, maintaining the landscape, and offering supplementary sources of food and income in times of economic difficulty (Fuller 1983). Rural advisory services have an important role to play to encourage part-time farmers to use their holdings in ways that minimize antagonism with full-time operators and to enable truly surplus land to be leased or otherwise be made available to neighbouring farmers (Krašovec 1983).

Promotion of farm-based tourism at a locally appropriate scale also offers important possibilities for generating additional income and for providing a means whereby urban visitors may learn about the realities of rural life and hence display more respect for the countryside (Dorfmann 1983).

OFF-FARM EMPLOYMENT

Many rural areas in the EC suffer disproportionately from high levels of unemployment, or from lack of opportunity for employment or decent income, or both. Many school leavers, women, unskilled workers and those aged 50 or more face particularly severe difficulties in their search for work within reasonable access of their home in the countryside; and seasonal unemployment raises a set of additional problems in holiday areas. Much more attention needs to be devoted to ensuring the provision of an adequate local range of non-agricultural employment in order to make use of the various elements contained in the rural labour pool, including young school-leavers with varying levels of education and skill, married women, ex-farmers and former agricultural workers. Generating adequate and fulfilling off-farm job opportunities for women is particularly vital if rural couples are to be 'anchored' in the countryside. Two inter-linked challenges are raised: namely to introduce completely new sources of employment (e.g. small 'nursery' factories, rural offices) and to support existing work-shops and other activities which may be on a precarious economic or management basis (Walker 1978). The work of the Development Commission in the UK and of the FIDAR in France provide many good examples of sensitive job creation in the countryside; and a new package of measures announced in the winter of 1982–3 for attempting to reconcile the 'development and protection of mountain areas' in France offers new opportunities for initiatives to be taken with a measure of financial support from government sources.

Alternative forms of employment need to be inserted in the countryside at a scale that will minimize damage to the environment and disruption to the social structure (Hodge and Whitby 1981). Small- or medium-scale solutions are normally appropriate but it is obvious that viable non-agricultural employment cannot be guaranteed in every rural settlement in the Community. Already there is a clear tendency for small factories to locate in small towns or key settlements where they can draw on the labour market of a sizeable commuting hinterland (Moseley 1979). Small craft industries and enterprises processing data rather than manufacturing products might be encouraged in villages and smaller settlements; however, good tele-communication links would have to be guaranteed (Clark and Unwin 1981). Creation and sustenance of small holding points merits considerable support, by virtue of the benefits they can impart to their immediate hinterlands and because rural workers' daily travel costs may be kept under control. However, it must be recognized that such small holding points will always experience strong competition from larger centres which command greater concentrations of economic activity and ranges of social facilities. To advocate job creation in the countryside is unquestionably a difficult case during the present time of economic difficulty when jobs are in short supply everywhere, but it is none the less essential if an active, off-farm population is to be retained in rural areas beyond the expanding outreach of regional cities.

RURAL SERVICES

Abundant scope exists for devising new approaches to service provision of all kinds in rural areas with low densities of population where levels of demand fall below normally accepted thresholds. All too often, aspects of organization and allocation of responsibilities that may be effective in urban areas prove to be out of scale and inappropriate in median and remote stretches of countryside. There is not a single, universal answer to this dilemma; instead, the challenge has to be met in various ways involving permutations of personal mobility, novel forms of 'public' transport, flexible opening hours (of shops, surgeries and the like), and implementation of innovative settlement policies in remote rural areas. The degree of official support for such experiments is highly varied and, in general terms, needs to be strengthened since it seems unquestionable that individual villages and groups of settlements will need to exercise greater grassroots initiative in the future. The idea of 'multiple functions' has much to commend it, whereby a local public official would perform a number of functions – including those traditionally provided by the post office – which normally required a range of civil servants to supply. Such an official might operate from a single place or, more imaginatively, might rotate between a number of settlements on specified days or parts of the week (Taylor and Emerson 1981). A multi-functional, non-specialist approach might also be advocated for the provision of basic care in rural areas, with an individual combining the roles of district nurse, health visitor and social worker. Serious cases of illness would, of course, require removal to the nearest hospital. Local voluntary organizations, with encouragement and modest financial support from official bodies, are often particularly appropriate for providing meals on wheels and other kinds of practical help to old people in the countryside. For example, in

Lower Saxony there are successful services known as *Sozialstationen* within which churches and local organizations operate, with the backing of local and provincial government, to provide various forms of assistance for elderly people in villages and more remote rural settlements.

Provision of local education for young children might be viewed more sympathetically by policy makers, rather than favouring the trend to close small village schools and take pupils by bus to centralized facilities. Special efforts would be required to attract qualified and devoted teachers to village schools to ensure that high educational standards be maintained. In addition there are other possibilities, including the management of rural schools in clusters that would share equipment and be served by peripatetic head teachers and specialist staff on specified days of the week, and involving parents to a much greater degree in the life of the school according to the principles of community schooling (Rogers 1979). An advance notice of impending closure would give parents and teachers time to investigate the possibility of an alternative solution and might also be used as a lever for residents to mount a campaign for additional housing or new employment facilities in an attempt to revitalize their threatened settlement. Considerable progress has already been made along these lines in many parts of rural France. The role and potential utility of the school as a community focus deserves greater recognition, while the use of buildings for pre-school playgroups, village meetings and for adult activities at evenings and weekends should be appreciated as one way of defraying some of the costs of keeping a village school open. However, it would be quite unrealistic to advocate that every tiny school could or should be reprieved (Comber 1981). Children of secondary-school age would have to continue to travel to towns or key settlements for more specialized training. Already some teenage pupils in rural Britain are involved in round trips of more than 100 km by bus each day. In such cases the burden of travel costs should not be allowed to fall on their parents but should be met from public funds.

Some sort of public transport will remain necessary, even in thinly populated rural areas, since a small but not insignificant residual demand remains. The car-less elderly, young, sick, and less affluent form the 'transport poor' who suffer particularly when rural services of many kinds are withdrawn by the various official and commercial concerns that ensured their provision in the past. Unconventional means of transport may well be appropriate for meeting that kind of need, rather than large public service vehicles that follow inflexible, timetabled routes. Various solutions are working successfully in different parts of the Community and there is plenty of scope for increasing their number. For example, post buses which carry passengers as well as mail have long operated in West Germany and parts of Scotland and have recently been introduced in several areas of rural England. Community minibuses, dial-a-ride services and voluntary car schemes all offer considerable potential but accommodation of adult passengers on school buses is less satisfactory. It is desirable that administrative, legal or financial hindrances to any kind of local initiative should be removed wherever possible so that rural residents may operate self-help schemes for service provision in their own localities.

Similarly in settlements where village shops have been forced to close, advice, encouragement and even limited financial support should be made available to help rural residents set up their own co-operative arrangements for purchasing and distributing food and other goods. Special 'rural officers' or 'field workers' can play a vital role in assisting self-help schemes which fit into a broader vision of a more humane brand of countryside management embracing concern for people living in rural areas and for conserving vulnerable environments. A case can be made for a greater number of rural officers to be trained and employed to operate co-ordinated advisory services in small rural districts. The work of the staff of rural community councils in the UK is exemplary in this respect.

If the Community's countryside is to retain some element of social mix it is essential that a variety of types of housing should be available. In particular, measures should be introduced to provide local workers with accommodation at a reasonable rent. Carefully sited social housing needs to be constructed for rural people who cannot cope with the marked inflation of house prices to which the arrival of retired people, second-home purchasers and other incomers makes a major contribution (Phillips and Williams 1982a, 1982b). Homes for young local couples should be provided as a priority and special finance should be devoted to this task.

In short, a strong argument may be offered for taking a range of initiatives to encourage rural residents and planners to co-operate in managing the Community's countryside in innovative and appropriate ways which embrace alternatives to established urban-inspired practice. In the language of development studies, rather more tolerance should be expressed toward endogenous 'bottom-up' ideas regarding rural management, rather than favouring 'top-down' schemes that seek to promote neat uniformity and are devised by remote administrators and politicians with limited knowledge of local conditions or the precise and diverse needs of people living in specific rural localities (Tracy 1982b).

The idea of a European rural fund has been advocated by Faure, Plumb and others and has found favour with the European Parliament and the Council of Europe, while special small sources of finance for imaginative rural projects are already in existence in some member states. The proposed fund would be modest in size, might well be supported as part of the Community's overall 'agricultural' budget (whatever its size) and might provide low-interest loans for:

stimulating and supporting small rural industries and other new sources of off-farm employment;

supporting rural services and facilities (especially those of an innovative or experimental nature);

enhancing the stock of low-cost housing for rural families by new construction and by renovation of delapidated but still habitable buildings;

encouraging job-creation schemes for rural women; and

assisting certain forms of agricultural activity, such as those that are geared to high-quality production using traditional methods or involve part-time farmers.

A further desirable means for encouraging new enterprises to develop in rural areas and vulnerable existing ones to survive would be to authorize some kind of exemption from taxation over a specified number of years in order to cushion delicate firms during particularly difficult phases of their existence. French experience of this kind of policy has been encouraging. As Tracy (1982b) has argued, large centrally conceived programmes may not be the most appropriate or effective for meeting the diverse needs of rural areas. Relatively inexpensive, inconspicuous but imaginative measures can often play a very important role in mobilizing human potential in the countryside.

Medium-technology agriculture may arguably be more appropriate in some parts of median and depopulated rural areas, but it remains clear that modern intensive farming will continue to advance in many dynamic sections of the countryside (Mendras 1979). The European

Commission should take a lead and offer its support to measures for monitoring the impact of agricultural chemicals, fertilizers and pesticides on wildlife and soil quality in both the short term and the longer perspective, and provide expert advice so that applications with ultimately unwise implications might be avoided. Intensification and conservation appear to be contradictions in terms and some kind of voluntary zoning agreement may be the only reasonable form of compromise although that is not beyond reproach and is fraught with practical difficulties.

A long-term and positive view of the countryside and its resources is absolutely essential for the well-being of the Community's residents – rural and urban alike – and this must be advocated in preference to any approach that concentrates on short-term expediency or emphasizes expenditure cuts to the exclusion of everything else (Clout 1982). Strategies must be based on a rounded, global understanding of the many roles of the countryside, of ecological principles of resource use, and of the economic and social needs of country people, rather than on the dangerous sectoral kind of approach whereby different agencies in every country of the Ten have pursued their own independent course and narrow, compartmentalized common policies have been devised which impinge on rural areas of the Community as a whole (Tracy 1982b). Hard thought must be directed to devising coherent rural policies which embrace matters that may appear to be discrete but in reality are strongly interrelated, such as settlement planning, service provision, methods of transport, use of energy, conservation of nature, techniques of agricultural activity, and generation of off-farm employment (O'Riordan 1983). There must be an official willingness to innovate and experiment in devising policies and to encourage the long-established but sometimes stifled capacity of residents of rural settlements to help themselves. With all urgency, politicians and policy-makers in the Community and the member states should look beyond their conventional sectoral responsibilities and start to think imaginatively and ecologically about the challenge of managing rural areas in their entirety, ensuring the well-being of their residents and encouraging the wise stewardship of vulnerable resources for use and enjoyment by generations of Europeans not yet born.

Postscript

The future of the CAP and the European countryside formed a major focus of concern in the final months of 1983. Mr Gaston Thorn, President of the European Commission, insisted that the effort to rationalize agriculture and to balance supply and demand should be accompanied by a parallel campaign to make the most of the countryside, not only from the point of view of landscape and protection of the environment, but also with respect to the development of other activities related to the land. Among these are the use of agricultural substances as sources of organic chemicals through biotechnology, materials to produce energy (biomass), and timber production. The EC has defects of both energy and timber but there are real and substantial possibilities for alternative activities and employment in rural areas. By combining remodelled agricultural activities with other improvements, he argued that it should be possible to raise the level of economic and social life in wide areas of the Community, without increasing the rural-urban exodus which had probably already reached the maximum desirable level.

Support for small farmers, increased development aid for poor rural areas, an improved division of available resources and control of food surpluses were the leading objectives of the European Commission's controversial proposals announced in the autumn of 1983 in order to reorganize the structure of farming in the EC. According to agriculture commissioner Poul Dalsager, such a policy would be vital for the Community's future. These proposals stem from decisions taken by the European Council in Stuttgart and represent part of a broad attempt to revitalize the EC. They are planned to remodel farming structures in the light of recent experience and with regard to changing market conditions for food products. Small farmers, who have suffered particularly in the recession and would be affected by a tighter farm budget and lower guaranteed commodity prices in the future, stood to gain most. It was proposed that they would receive additional aid, with the Commission's programme relaxing conditions for assistance in order to help less prosperous farmers, retain more people in land-based employment, and, at the same time, improve land management. Greece, Italy and the Republic of Ireland would receive major investments under the plan, which would offer further financial aid to compensate for 'natural handicaps' and infrastructural problems in backward rural areas and to support a large number of local integrated schemes. In the past, subsidies and investments were strongly geared to 'development' farms and to projects that would increase productivity. Recommended new priorities include reducing production costs, improving the living and working conditions of the rural population, and enhancing the quality of food produced. In addition, the proposals would allow a reduction in commodity 'lakes' and 'mountains' by encouraging the processing of surplus goods into saleable products. If implemented, the proposals would represent only a timid step along the path toward a rural policy for the EC and planners and policy makers need to be constantly attentive to the ecological implications of the ideas that they propose.

Early in the long-awaited year of 1984 European politicians continue to fight skirmishes and whole battles over the future size of the Community budget, over acceptable national contributions to it, and over the desirable magnitude that the new CAP should occupy within it. With remarkable directness, Mr Dalsager has argued that the CAP – in its established form – is a luxury that the EC can no longer afford. The farm policy needs to be lean and hungry if it is to survive. He warned that should the CAP collapse, then 'the whole edifice of the European Community will be at the brink of failure'. As well as demanding a redefinition of financial support, the slimmed-down CAP that is proposed for the future could usefully be recast to embrace rural matters as a whole rather than concentrating overwhelmingly on agricultural prices and production. Many would argue that the time for such a change is long overdue.

References

Aalen, F. H. A. (1978) *Man and the Land in Ireland*, London, Academic Press.

Abercrombie, K. (1981) *Rural Development in Lewis and Harris: the Western Isles of Scotland*, Langholm, The Arkleton Trust.

Abrami, A. (1978) 'Italian legislation on mountain land and on uncultivated or insufficiently cultivated land', *Landscape Planning*, 5, 171-9.

Adams, W. (1983) 'Conflict in the Western Isles', *Geographical Magazine*, 54, 340-1.

Albarre, G. (1982) 'La rénovation rurale', *Ruralités Nouvelles*, 6, 1-32.

Albarre, G. and Sottiaux, J. P. (1981) 'Transports en régions rurales', *Ruralités Nouvelles*, 2, 1-36.

Alexander, D. E. (1980) 'I Calanchi: accelerated erosion in Italy', *Geography*, 65, 95-100.

Allardt, E. (1982) 'Reflections on the rural nature of past and present', *Sociologia Ruralis*, 22, 99-107.

Anon.
(1973) *Second Report from the Select Committee of the House of Lords on Sport and Leisure*, London, HMSO.
(1980a) 'L'évolution des zones rurales wallonnes', *Ruralités Nouvelles*, 1, 1-32.
(1980b) 'La politique montagne', *Le Mois de l'Environnement*, 5, 1-28.
(1982) 'Scottish Highlands: life in the outback', *The Economist*, 24 July, 25.

Ardagh, J. (1982) *France in the 1980s*, Harmondsworth, Penguin.

Arkleton Trust (1982) *Schemes of Assistance to Farmers in Less-Favoured Areas of the EEC*, Langholm.

Aubert, B. (1971) 'Parc national des Pyrénées occidentales et sa zone périphérique', in J. Kostrowicki and M. Rosciszewski (eds)

L'Aménagement de la Montagne, Warsaw, Państwowe Wydawnictwo Naukowe, 183–95.

Auriac, F. and Bernard, M. C. (1972) 'Les ouvriers agricolesexploitants à temps partiel en Languedoc-Roussillon', in *Actes du Colloque de Géographie Agraire,* Aix-en-Provence, Institut de Géographie, 107–12.

Bagnaresi, U. (1978) 'An approach to the planning of abandoned farmland in Italy', *Landscape Planning,* 5, 157–69.

Balsan, L. (1973) *Larzac, Terre Méconnue,* Paris, Editions Ouvrières.

Barberis, C.
(1968) 'The agricultural exodus in Italy', *Sociologia Ruralis,* 8, 179–88.
(1973) 'Les ouvriers-paysans en Europe et dans le monde', *Etudes Rurales,* 49–50, 106–21.

Barbichon, G. (1973) 'Appropriation urbaine du milieu rural à des fins de loisirs', *Etudes Rurales,* 49–50, 97–105.

Barbier, B. (1972) 'Le rôle des petites villes en milieu montagnard', *Bulletin de l'Association de Géographes Français,* 400–1, 295–8.

Barthélemy, D. and Barthez, A. (1978) 'Propriété foncière, exploitation agricole et aménagement de l'espace rural', *Economie Rurale,* 126, 6–16.

Bauer, G. and Roux, J. M. (1976) *La Rurbanisation ou la Ville Eparpillée,* Paris, Seuil.

Belliard, J. C. and Boyer, J. C. (1983) 'Les "nouveaux ruraux" en Ilede-France', *Annales de Géographie,* 92, 433–51.

Berger, M. and Fruit, J. P. (1980) 'Rurbanisation et analyse des espaces ruraux péri-urbains', *L'Espace Géographique,* 9, 303–13.

Best, R. H.
(1979) 'Land-use structure and change in the EEC', *Town Planning Review,* 50, 395–411.
(1981) *Land Use and Living Space,* London, Methuen.

Béteille, R.
(1977) 'Points d'ancrage en milieu rural', *Norois,* 24, 501–24.
(1981) *La France du Vide,* Paris, Librairies Techniques.

Bethemont, J. (1977) 'L'irrigation en Espagne', *Revue Géographique des Pyrénées et du Sud-Ouest,* 48, 357–86.

Bethemont, J. and Pelletier, J. (1983) *Italy: a Geographical Introduction,* London, Longman.

Beynon, J. (1979) 'Dutch treats', *Town and Country Planning,* 48, 88–90.

Billet, J. (1976) 'The mountain regions of Europe', *Nateuropa*, 25, 8–10.

Blacksell, M. (1976) 'The role of le parc naturel et régional', *Town and Country Planning*, 44, 165–70.

Blotevogel, H. H. and Hommel, M. (1980) 'Current trends in the development of urban systems', *Bochumer Geographische Arbeiter*, 38, 25–33.

Body, R. (1982) *Agriculture: the Triumph and the Shame*, London, Temple Smith.

Boichard, J. (1958) 'Le niveau de vie du paysan', *Revue de Géographie de Lyon*, 33, 25–55.

Bonnamour, J. (1973) *Géographie Rurale: Méthodes et Perspectives*, Paris, Masson.

Bonnier, J. and Coste, M. (1978) 'Consommation d'espaces et habitat individuel', *Revue de Géographie de Lyon*, 53, 313–37.

Bord na Gaelige (1983) *Action Plan for Irish 1983-1986*, Dublin.

Bouet, G. and Fel, A. (1983) *Le Massif Central*, Paris, Flammarion.

Bouquet, M. (1982) 'Production and reproduction of family farms in south-west England, *Sociologia Ruralis*, 22, 227–44.

Boutet, R. (1973) 'La France pauvre, ou l'autre France', *Travaux de l'Institut de Géographie de Reims*, 16, 57–70.

Bowler, I. (1976) 'Recent developments in the agricultural policy of the EEC', *Geography*, 61, 28–30.

Bradley, C. (1982) 'Historical background to the crofting problem', *Ecos*, 3, 21–3.

Bravard, J. P. (1975) 'Les citadins et la forêt: l'exemple du Parc du Pilat', *Revue de Géographie de Lyon*, 50, 151–65.

British Trust for Ornithology (1979) *Atlas of Breeding Birds*, London.

Brody, H. (1974) *Inishkillane: Change and Decline in the West of Ireland*, New York, Schocken.

Broggi, M. F. (1979) *Compatability of Agriculture and Forestry Activities and Protection of the Environment*, Strasbourg, Council of Europe.

Brotherton, I.
(1982a) 'Development pressures and control in the national parks 1966-81', *Town Planning Review*, 53, 439–59.

(1982b) 'National parks in Great Britain and the achievement of nature conservation purposes', *Biological Conservation*, 22, 85–100.

Brun, A. (1978) 'Concurrence agriculture-forêt en moyenne montagne', *Economie Rurale*, 127, 54–7.

Brunet, P. (1974) 'L'évolution récente des paysages ruraux français', *Geographia Polonica*, 29, 13–30.

Bryant, C. R., Russwurm, L. H. and McLellan, A. G. (1983) *The City's Countryside: Land and its Management in the Rural-urban Fringe*, London, Longman.

Bryden, J. (1981) 'Appraising a regional development programme: the case of the Scottish Highlands and Islands', *European Review of Agricultural Economics*, 8, 475–97.

Bryden, J. and Houston, G. (1976) *Agrarian Change in the Scottish Highlands: the Role of the HIDB in the Agricultural Economy of the Crofting Counties*, London, Martin Robertson.

Burdekin, D. A. (ed.) (1983) 'Research on Dutch elm disease in Europe', *Forestry Commission Bulletin*, 60, 1–114.

Burgel, G.

(1972) 'Recherches agraires en Grèce', *Mémoires et Documents du Centre National de la Recherche Scientifique*, 13, 7–62.

(1975) 'Unité et diversité des paysages agraires grecs', in Deputazione di Storia Patria per l'Umbria, *I Paesaggi Rurali Europei*, Perugia, 57–65.

(1981) *Croissance Urbaine et Développement Capitaliste. Le Miracle Athénien*, Paris, Centre Nationale de la Recherche Scientifique.

Buttel, F. H. (1980) 'Agricultural structure and rural ecology: toward a political economy of rural development', *Sociologia Ruralis*, 20, 44–62.

Caird, J. B.

(1972) 'Changes in the Highlands and Islands of Scotland', *Geoforum*, 12, 5–36.

(1975) 'Problems of the transformation of agricultural settlements: the crofting settlements of the Outer Hebrides, Scotland', in B. Saffalvi (ed.) *Urbanization in Europe*, Budapest, Akadémiai Kiadó, 279–94.

Calmès, R.

(1978) *L'Espace Rural Français*, Paris, Masson.

(1981) 'L'évolution des structures d'exploitation dans les pays de la CEE', *Annales de Géographie*, 90, 401–27.

Canevet, C. (1979) 'De la polyculture paysanne à l'intégration: les couches sociales dans l'agriculture', *Norois*, 26, 507–22.

Carbinier, R. (1978) 'Forests or just trees?', *Nateuropa*, 31, 10–13.

Cardol, G. and Van Engelenburg, R. F. C. (1980) 'Development of the regional imbalance in the European Community 1970-1977',

European Parliament, Research and Documentation Papers, Regional Policy and Transport Series, 11, 1–36.

Caron, J. (1975) 'Mode de vie du Roumois: l'agriculture à temps partiel', Economie Rurale, 110, 37–43.

Carrière, P.

(1973) 'Viticulture et espace rural', Bulletin de la Société Languedocienne de Géographie, 7, 221–39.

(1975) 'Aménagement agricole et évolution du paysage rural en Languedoc', in Deputazione di Storia Patria per l'Umbria, I Paesaggi Rurali Europei, Perugia, 67–83.

Carter, F. W. (1968) 'Population migration to greater Athens', Tijdschrift voor Economische en Sociale Geografie, 59, 100–5.

Cassola, F. and Lovari, S. (1976) 'Nature conservation in Italy', Biological Conservation, 9, 243–57.

Castela, P. (1976) 'L'évolution des rapports entre ville et campagne dans la région niçoise', Recherches Géographiques à Strasbourg, 77–86.

Cawley, M. E.

(1979) 'Rural industrialization and social change in western Ireland', Sociologia Ruralis, 19, 43–59.

(1983) 'Part-time farming in rural development: evidence from western Ireland', Sociologia Ruralis, 23, 63–75.

Chevalier, M. (1981) 'Les phénomènes néo-ruraux', L'Espace Géographique, 10, 33–47.

Chisholm, M. (1962) Rural Settlement and Land Use, London, Hutchinson.

Christaller, W.

(1933) Die Zentralen Orte in Süddeutschland, Jena.

(1964) 'Some considerations of tourism in Europe; the peripheral regions, underdeveloped countries, recreation areas', Papers of the Regional Science Association, 12, 95–105.

Christians, C.

(1977) 'Les régions défavorisées en Europe', Bulletin de la Société Géographique de Liège, 13, 207–39.

(1978) 'Pédologie, affectation du sol et aspects régionaux d'aménagement du territoire au Grand-Duché de Luxembourg', Bulletin de la Société Géographique de Liège, 14, 13–44.

(1979) 'L'évaluation des paysages et des sites ruraux: essais de méthode et résultats dans quelques régions wallonnes', Bulletin de la Société Géographique de Liège, 15, 167–208.

(1980) 'Les résultats de 25 années de modernisation d'une agriculture avanceé: l'exemple belge', *Hommes et Terres du Nord,* 23-40.

(1981) 'Législation et réalisation de l'aménagement rural en Belgique', *Demain: Revue de la Société d'Etudes et d'Expansion,* 289, 395-405.

(1982) 'Les types d'espaces ruraux en Belgique', *Hommes et Terres du Nord,* 16-28.

Christophe, G. (1982) 'Les services et le tourisme en milieu rural', in *Initiative Economique en Milieu Rural,* Arlon, Fondation Rurale de la Wallonie, 99-105.

Clark, D. and Unwin, K. I. (1981) 'Telecommunications and travel: potential impact in rural areas', *Regional Studies,* 15, 47-56.

Clark, G.

(1982a) 'Housing policy in the Lake District', *Transactions, Institute of British Geographers,* 7, 59-70.

(1982b) *Housing and Planning in the Countryside,* Chichester, Wiley.

Cléac'h, A. (1977) 'La basse vallée de l'Aulne: une région à vocation paysagère et agricole', *Norois,* 24, 525-41.

Clément, F. (1978) 'Approches de l'animation de développement micro-régional', *Pour,* 60, 50-61.

Cloke, P.

(1979) *Key Settlements in Rural Areas,* London, Methuen.

(1983) *An Introduction to Rural Settlement Planning,* London, Methuen.

Clout, H. D.

(1969a) 'Second homes in France', *Journal of the Town Planning Institute,* 55, 440-3.

(1969b) 'Problems of rural planning in the Auvergne', *Planning Outlook,* 6, 29-37.

(1970) 'Social aspects of second home occupation in the Auvergne', *Planning Outlook,* 9, 33-49.

(1972) 'Part-time farming in the Puy-de-Dôme département', *Geographical Review,* 62, 271-3.

(1974) 'Agricultural plot consolidation in the Auvergne', *Norsk Geografisk Tiddskrift,* 28, 181-94.

(1975) 'La belle France', *Geographical Magazine,* 47, 302-9.

(1976a) 'The growth of second home ownership: an example of seasonal suburbanization', in J. H. Johnson (ed.) *Suburban*

Growth, Chichester, Wiley, 101–27.

(1976b) 'Recreation in an urban society: national parks and second homes', in R. Lee and P. E. Ogden (eds) *Economy and Society in the EEC*, Farnborough, Saxon House, 127–49.

(1979) 'Land-use change in Finistère during the eighteenth and nineteenth centuries', *Etudes Rurales*, 73, 69–96.

(1982) 'The decline of regional economies in Europe in predominantly rural regions', *Institute of European Environmental Policy, Occasional Papers*, 1–26.

Cohou, M. (1977) 'La population non-agricole au village: différenciation et prolétarisation de la société rurale', *Etudes Rurales*, 67, 47–59.

Comber, L. C. (1981) *The Social Effects of Rural Primary School Reorganization in England*, Birmingham, University of Aston Joint Unit for Research on the Urban Environment.

Commins, P. (1980) 'Imbalances in agricultural modernization: with illustrations from Ireland', *Sociologia Ruralis*, 20, 68–81.

Commission of the European Communities (1981) *The Agricultural Situation in the Community*, Luxembourg, European Commission.

Connell, J. H. (1978) *The End of Tradition: Country Life in Central Surrey*, London, Routledge & Kegan Paul.

Coppock, J. T. (ed.) (1977) *Second Homes: Curse or Blessing?*, Oxford, Pergamon.

Council for the Protection of Rural England (1975) *Landscape: the Need for a Public Voice*, London.

Courgeau, D. and Lefebvre, M. (1982) 'Les migrations internes en France de 1954 à 1975: migrations et urbanisation', *Population*, 37, 341–70.

Cribier, F.

(1973) 'Les résidences secondaires des citadins dans les campagnes françaises', *Etudes Rurales*, 49–50, 181–204.

(1975) 'Retirement migration in France', in L. A. Kosiński and R. M. Prothero (eds) *People on the Move: Studies on Internal Migration*, London, Methuen, 361–73.

(1982) 'Aspects of retired migration from Paris', in A. M. Warnes (ed.) *Geographical Perspectives on the Elderly*, Chichester, Wiley.

Cros, Z. (1980) 'Fostering public awareness of the landscape during the preparation of a rural development plan', *Landscape Planning*, 7, 263–79.

Daudé, G.

(1973) 'Ecologie et humanisme à travers l'exemple du parc national des Cévennes', *Travaux de l'Institut de Géographie de Reims,* 14, 93–103.

(1976a) 'Une action du parc national des Cévennes: l'opération des hameaux', *Revue de Géographie de Lyon,* 51, 175–88.

(1976b) 'Les parcs naturels français', *Revue de Géographie de Lyon,* 51, 99–105.

Davidson, J. and Lloyd, R. (eds) (1977) *Conservation and Agriculture,* Chichester, Wiley.

De Angelis, M. E. S. and Patella, L. V. (1978) 'Reduction of agricultural land in Umbria 1970–1975', *Geographia Polonica,* 38, 11–18.

De Benedictis, M. (1981) 'Agricultural development in Italy: national problems in a Community framework', *Journal of Agricultural Economics,* 32, 275–85.

De Farcy, H. (1975) *L'Espace Rural,* Paris, Presses Universitaires de France.

Delamarre, A. (1976) 'Les bâtiments modernes d'élevage en France', *Revue Géographique des Pyrénées et du Sud-Ouest,* 47, 139–58.

Delbos, G. (1979) 'A l'ombre des usines: industrialisation et maintien d'une communauté paysanne en Lorraine', *Etudes Rurales,* 76, 83–96.

Dematteis, G. (1982) 'Repeuplement et revalorisation des espaces périphériques: le cas de l'Italie', *Revue Géographique des Pyrénées et du Sud-Ouest,* 53, 129–43.

Deniel, J. (1965) 'Les talus et l'aménagement de l'espace rural', *Penn Ar Bed,* 12, 41–54.

De Réparaz, A.

(1982a) 'La montagne provençale investie: les nouveaux pouvoirs et les agents extérieurs de la transformation contemporaine dans la Haute-Provence montagnarde', *Provence Historique.* 32, 391–407.

(1982b) *Nouvel Atlas Rural de la Région Provence-Alpes-Côte d'Azur,* Université d'Aix, Aix-en-Provence.

De Réparaz, A. and Bernard, C. (1975) 'Viticulture et coopération vinicole dans le sud-est méditerranéen', *Méditerranée,* 4, 37–57.

Desplanques, H.

(1957) 'La réforme agraire italienne', *Annales de Géographie,* 66, 310–27.

(1973) 'Une nouvelle utilisation de l'espace rural en Italie:

l'agritourisme', *Annales de Géographie*, 82, 151–64.

Dessau, J. (1982) 'Another agriculture?', *Sociologia Ruralis*, 22, 167–71.

De Veer, A. A. and Burrough, P. A. (1978) 'Physiognomic landscape mapping in the Netherlands', *Landscape Planning*, 5, 45–62.

Dewailly, J. M. and Dubois, J. J. (1977) 'Necessité et ambiguité récréative de la forêt', *Hommes et Terres du Nord*, 5–30.

Dewhurst, J. F., Coppock, J. O. and Yates, P. L. (1961) *Europe's Needs and Resources*, London, Macmillan.

Dickinson, R. E.
(1954) 'Land reform in southern Italy', *Economic Geography*, 30, 157–76.
(1957) 'The geography of commuting: the Netherlands and Belgium', *Geographical Review*, 47, 521–8.
(1959) 'The geography of commuting in West Germany', *Annals of the Association of American Geographers*, 49, 443–56.

Dierschke, H. (1978) 'Monotony: grassland vegetation', *Nateuropa*, 31, 29–32.

Dietrich, G. (1976) 'Forêts communales et communautés rurales dans le département des Vosges', *Revue Géographique de l'Est*, 16, 41–61.

Diry, J. P. (1975) 'L'innovation en milieu rural: l'élevage industriel', *L'Espace Géographique*, 4, 287–93.

Dorfmann, M. (1983) 'Régions de montagne: de la dépendance à l'auto-développement?', *Revue de Géographie Alpine*, 71, 5–34.

Douguédroit, A. (1981) 'Reafforestation in the French southern Alps', *Mountain Research and Development*, 1, 245–52.

Drewett, R. (1980) 'Changing urban structures in Europe', *Annals of the American Academy of Political and Social Science*, 451, 52–75.

Drudy, P. J. (1978) 'Depopulation in a prosperous agricultural subregion', *Regional Studies*, 12, 49–60.

Duboscq, P. (1976) 'Les paysans et leur logement dans le sud-ouest aquitain', *Revue Géographique des Pyrénées et du Sud-Ouest*, 47, 121–38.

Duffey, E. (1982) *National Parks and Reserves of Western Europe*, London, Macdonald.

Dunn, M. C., Rawson, M. and Rogers, A. (1981) *Rural Housing: Competition and Choice*, London, Allen & Unwin.

EC
(1980) 'Motion for a resolution for a new European agricultural policy, submitted to the Committee on Agriculture', *European*

Parliament: Working Documents, 1–800/79, Luxembourg.

(1981a) 'Motion for a resolution on possible improvements to the Common Agricultural Policy', *European Parliament: Working Documents,* 1–250/81, Luxembourg.

(1981b) 'Report drawn up on behalf of the Committee on Regional Policy and Regional Planning on the contribution of rural development to the re-establishment of regional balances in the Community', *European Parliament: Working Documents,* 1–648/81, Luxembourg.

(1982) 'Opinion of the Committee on Agriculture on the contribution of rural development to the re-establishment of regional balances in the Community', *European Parliament: Working Documents,* 1–648/81/Annex, Luxembourg.

Edwards, R. (1983) 'Scotland: behind the scenery', *Social Work Today,* 14(27), 10–14.

Ellington, A. and Burke, T. (1981) *Europe: Environment,* London, Ecobooks.

Enyedi, G. (1975) *Research Problems in Rural Geography,* Budapest, International Geographical Union.

Fairhall, J. (1971) 'Farming Europe's landscapes', *Nature in Focus,* 10, 16–19.

Fel, A. (1974) 'Types d'évolution démographique de la France rurale', *Geographia Polonica,* 29, 55–62.

Fel, A. and Miège, J. (1972) 'Transformation et urbanisation des campagnes en Allemagne fédérale', *Annales de Géographie,* 31, 579–93.

Fennell, R. (1981) 'Farm succession in the European Community', *Sociologia Ruralis,* 21, 19–42.

Ferras, R., Picheral, H. and Vielzeuf, B. (1979) *Languedoc et Roussillon,* Paris, Flammarion.

Ferré, P. (1981) 'Aménagement volontaire et essai de décentralisation', *Bulletin de la Société Languedocienne de Géographie,* 15, 339–69.

Fielding, A. J. (1982) 'Counterurbanization in Western Europe', *Progress in Planning,* 17, 1–52.

Firnberg, H. (1971) 'Transfrontier natural parks', *Nature in Focus,* 11-2-4.

Fitton, M. (1976) 'The urban fringe and the less privileged', *Countryside Recreation Review,* 1, 25–34.

Flatrès, P. (1979) 'L'évolution des bocages: la région de Bretagne', *Norois,* 26, 303–20.

Flatrès-Mury, H. (1970) 'Matériaux et techniques de construction rurale dans l'ouest de la France', *Norois*, 17, 547–65.

Fornier, J. W. (1979) 'Managing the natural environment in agricultural areas', *Planning and Development in the Netherlands*, 11, 161–78.

Forsythe, D. E. (1980) 'Urban incomers and rural change: the impact of migrants from the city on life in an Orkney community', *Sociologia Ruralis*, 20, 287–307.

Frank, W. (1983) 'Part-time farming, underemployment and double activity of farmers in the EEC', *Sociologia Ruralis*, 23, 20–7.

Franklin, S. H.
    (1961) 'Social structure and land reform in southern Italy', *Sociological Review*, 9, 323–49.
    (1964) 'Gosheim, Baden-Württemberg: a Mercedes Dorf', *Pacific Viewpoint*, 5, 127–58.
    (1969) *The European Peasantry: the Final Phase*, London, Methuen.
    (1971) *Rural Societies*, London, Macmillan.

Frelastre, G. (1974) 'Le temps et la concertation, facteurs importants de l'aménagement du territoire: les Landes de Gascogne', *Economie Rurale*, 101, 33–8.

Fruit, J. P. (1983) 'L'agriculture dans les pays de la CEE', *L'Information Géographique*, 47, 66–7.

Fuguitt, G. (1977) 'Part-time farming: its nature and implications', *Centre for European Agricultural Studies, Wye College, Seminar Papers*, 2, 1–80.

Fuller, A. M. (1983) 'Part-time farming and the farm family', *Sociologia Ruralis*, 23, 5–10.

Gade, D. W. (1978) 'Windbreaks in the lower Rhône valley', *Geographical Review*, 68, 127–44.

Gasson, R.
    (1966) 'The influence of urbanization on farm ownership and practice', *Studies in Rural Land Use, Wye College*, 7.
    (1982) 'Part-time farming in Britain: research in progress', *GeoJournal*, 6, 355–8.

Gay, F. and Wagret, P. (1979) *L'Economie de l'Italie*, Paris, Presses Universitaires de France.

Gentilcore, R. L. (1970) 'Reclamation in the Agro Pontino, Italy', *Geographical Review*, 60, 301–27.

George, P. (1949) *Géographie Economique et Sociale de la France*, Paris, Editeurs Réunis.

Gervais, M., Servolin, C. and Weil, J. (1965) *Une France sans Paysans,* Paris, Seuil.

Gilbert, Y. (1978) 'Le mythe rural', *Espaces et Sociétés,* 24–7, 3–27.

Gillmor, D. A. (1977) 'EEC scheme for farming in less favoured areas', *Irish Geography,* 10, 101–8.

Gillon, S. (1981) 'Selling rural council houses', *Town and Country Planning,* 50, 115–16.

Gojceta, D. (1978) 'Les résidences secondaires dans les bassins de l'Aisne et de la Lienne', *Bulletin de la Société Géographique de Liège,* 14, 13–44.

Gold, J. R. and Burgess, J. A. (eds) (1982) *Valued Environments,* London, Allen & Unwin.

Grainger, A. (1981) 'Reforesting Britain', *The Ecologist,* 11, 56–81.

Green, B. H.
(1975) 'The future of the British countryside', *Landscape Planning,* 2, 179–95.
(1980) *Countryside Conservation,* London, Allen & Unwin.

Grosso, R. (1973) 'Le renouveau villageois sur la rive gauche du Rhône entre Drôme et Durance', *Etudes Rurales,* 49–50, 265–95.

Guellec, A. (1979) 'Les types de remembrement dans le centre Bretagne', *Norois,* 26, 404–7.

Guermond, Y. (1973) 'L'espace rural face à l'urbanisation', *Cahiers Géographiques de Rouen,* 1, 3–63.

Hall, P. G. (1980) 'New trends in European urbanization', *Annals of the American Academy of Political and Social Science,* 451, 45–51.

Hall, P. G. and Hay, D. (1980) *Growth Centres in the European Urban System,* London, Heinemann.

Hannan, D. F.
(1969) 'Migration motives and migration differentials among Irish rural youth', *Sociologia Ruralis,* 9, 195–220.
(1970) *Rural Exodus,* London, Chapman.

Hanrahan, P. J. and Cloke, P. J. (1983) 'Towards a critical appraisal of rural settlement planning in England and Wales', *Sociologia Ruralis,* 23, 109–29.

Hardy, P. and Matthews, R. (1977) 'Farmland tree survey of Norfolk', *Countryside Recreation Review,* 2, 31–3.

Harms, W. B. (1982) *Effects of Intensification of Agriculture on Nature and Landscape in the Netherlands,* Wageningen, University of Wageningen.

Hartke, W. (1956) 'Die Sozialbrache als Phänomen der geographischen Differenzierung der Landschaft', *Erdkunde*, 10, 257–69.

Harvois, P. (1978) 'Le mileu rural', *Pour*, 60, 26–30.

Hays, D. (1976) 'Le parc naturel régional des Landes de Gascogne', *Revue Géographique des Pyrénées et du Sud-Ouest*, 47, 381–98.

Hélias, P. J. (1975) *Le Cheval d'Orgueil*, Paris, Plon.

Higgs, J. (1983) *Institutional Approaches to Rural Development in Europe*, Langholm, The Arkleton Trust.

Hochas, J. (1977) 'Aménager le territoire: avec quels hommes?', *Economie Rurale*, 118, 14–23.

Hodge, I. and Whitby, M. (1981) *Rural Employment: Trends, Options, Choices*, London, Methuen.

Holmes, G. M. (1973) 'Slices of Danish land', *Geographical Magazine*, 45, 772–3.

House of Lords, Select Committee on the European Communities (1980) *Policies for Rural Areas in the European Community, Session 1979-80, 27th Report*, London, HMSO.

Houston, J. M. (1964) *The Western Mediterranean World*, London, Longman.

Hugonie, G. (1976) 'L'évolution récente de l'utilisation des sols montagnards en Sicile septentrionale', *Méditerranée*, 24, 3–17.

International Labour Office (1971) 'The social situation of old people in rural areas', *Sociologia Ruralis*, 11, 374–400.

Jansen, A. J.
  (1969) 'Social implications of farm mechanization', *Sociologia Ruralis*, 9, 340–407.
  (1974) 'Explorations into the future of agriculture in Western Europe', *Sociologia Ruralis*, 14, 45–53.

Joannon, M. (1975) 'Le village dans les campagnes provençales: analyse de l'évolution récente des villages perchés', in Deputazione di Storia Patria per l'Umbria, *I Paesaggi Rurali Europei*, Perugia, 303–16.

Johnson, J. H. (ed.) (1976) *Suburban Growth*, Chichester, Wiley.

Johnson, M. (1979) 'The co-operative movement in the Gaeltacht', *Irish Geography*, 12, 68–81.

Jones, P. (1981) 'The geography of the Dutch elm disease in Britain', *Transactions, Institute of British Geographers*, 6, 324–36.

Jordan, T. (1973) *The European Culture Area*, London, Harper International.

Juillard, E. (1973) 'Urbanisation des campagnes', *Etudes Rurales*, 49–50, 5–9.

Kalaora, B. (1978) 'L'ordre et la nature: le vert endimanché', *Espaces et*

Sociétés, 24–7, 29–37.

Kalaora, B. and Pelosse, V. (1977) 'La forêt loisir, un equipement de pouvoir', *Hérodote*, 7, 92–129.

Kariel, H. G. and P. E. (1982) 'Socio-cultural impacts of tourism: an example from the Austrian Alps', *Geografiska Annaler*, 64B, 1–16.

Kayser, B.
(1969) 'L'espace non-métropolisé du territoire français', *Revue Géographique des Pyrénées et du Sud-Ouest*, 40, 371–8.
(1972) 'El espacio rural y el nuevo sistema de relaciones cuidad-campo', *Revista de Geografía*, 6, 209–17.
(1973) 'Le nouveau système des relations villes-campagnes', *Espaces et Sociétés*, 8, 3–13.
(1980) 'Le changement social dans les campagnes françaises', *Economie Rurale*, 135, 1–7.

Kayser, B. and Thompson, K. (eds) (1964) *Economic and Social Atlas of Greece*, Athens, National Statistical Service of Greece.

Kearns, K. (1974) 'Resuscitation of the Irish Gaeltacht', *Geographical Review*, 64, 82–110.

King, R. L. (1973) *Land Reform: the Italian Experience*, London, Butterworth.

King, R. L. and Strachan, A. (1978) 'Sicilian agro-towns', *Erdkunde*, 32, 110–23.

King, R. L. and Young, S. (1979) 'The Aeolian islands: birth and death of a human landscape', *Erdkunde*, 33, 193–204.

Klöpper, R.
(1971) 'The urbanization of rural districts in Western Germany', in F. Dussart (ed.) *L'Habitat et les Paysages Ruraux de l'Europe*, Liège, Université de Liège, 283–91.
(1976) 'Federal Republic of Germany', in Economic Commission for Europe, *Planning and Development of the Tourist Industry in the EEC Region*, New York, United Nations, 51–60.

Kormoss, I. B. F. (1976) 'Urban extensions in Belgium', in P. Laconte (ed.) *The Environment of Human Settlements*, Oxford, Pergamon.

Kosiński, L. A. (1970) *The Population of Europe*, London, Longman.

Kouloussi, F. (1983) 'L'emploi dans l'agriculture des neuf pays de la CEE', *Revue du Marché Commun*, 262, 12–25.

Krašovec, S. (1983) 'Farmers' adjustment to pluriactivity', *Sociologia Ruralis*, 23, 11–19.

Küpper, U. I. (1969) 'Socio-geographical impacts of industrial growth at Shannon', *Irish Geography*, 6, 14–29.

Lacoste, Y. (1977) 'A quoi sert le paysage?', *Hérodote*, 7, 3–41.

Laganier, C. (1977) 'Retraités et résidents secondaires en Basse-Ardèche', *Bulletin de la Société Languedocienne de Géographie*, 11, 141–51.

Lambert, A. M.
(1961) 'Farm consolidation and improvement in the Netherlands', *Economic Geography*, 37, 115–23.
(1963) 'Farm consolidation in Western Europe', *Geography*, 48, 31–48.
(1971) *The Making of the Dutch Landscape*, London, Seminar Press.

Larrère, G. R. (1978) 'Désertification ou annexion de l'espace rural? L'exemple du plateau de Millevaches', *Etudes Rurales*, 71–2, 9–48.

Laurent, C.
(1975) 'Quinze ans d'évolution des populations rurales françaises', *Economie Rurale*, 105, 3–7.
(1982) 'Multiple jobholding farmers in agricultural policy', *GeoJournal*, 6, 287–92.

Law, C. M. and Warnes, A. M. (1980) 'The characteristics of retired migrants', in D. T. Herbert and R. J. Johnston (eds) *Geography and the Urban Environment 3*, Chichester, Wiley.

Leardi, E. (1973) 'L'apiezza demografica dei comuni italiani: i comuni minimi', *Bollettino della Società Geografica Italiana*, 10(2), 341–72.

Lebeau, R. (1969) *Les Grands Types de Structures Agraires dans le Monde*, Paris, Masson.

Le Coz, J. (1973) 'Niveaux de structuration de l'espace rural français', *Bulletin de la Société Languedocienne de Géographie*, 7, 135–68.

Léger, D. and Hervieu, B. (1979) *Le Retour à la Nature*, Paris, Seuil.

Leonard, P. L. and Cobham, R. O. (1977) 'The farming landscapes of England and Wales: a changing scene', *Landscape Planning*, 4, 205–36.

Leroy, C. M. and Probst, G. (1982) 'La rénovation du village en Allemagne', *Ruralités Nouvelles*, 5, 1–32.

Leroy, L. (1960) *Le Ruralisme*, Paris, Editions Ouvrières.

Lewis, G. J. and Maund, D. J. (1976) 'The urbanization of the countryside: a framework for analysis', *Geografiska Annaler*, 58B, 17–27.

Lichtenberger, E.
(1975) *The Eastern Alps*, London, Oxford University Press.
(1978) 'The crisis of rural settlement and farming in the high mountain region of continental Europe', *Geographia Polonica*, 38, 181–7.

Lieutaud, J. (1982) 'Industrie de base et espace économique: le complexe sidérurgique de Tarente et son impact local', *Annales de Géographie*, 91, 551-72.

Limouzin, P. (1980) 'Les facteurs de dynamisme des communes rurales françaises', *Annales de Géographie*, 89, 549-87.

Livet, R. (1965) *L'Avenir des Régions Agricoles*, Paris, Editions Ouvrières.

Lovari, S. and Cassola, F. (1975) 'Nature conservation in Italy: the existing national parks and other protected areas', *Biological Conservation*, 8, 127-42.

Lowe, P. and Goyder, J. (1983) *Environmental Groups in Politics*, London, Allen & Unwin.

McDermott, D. and Horner, A. (1978) 'Aspects of rural renewal in western Connemara', *Irish Geography*, 11, 176-9.

McEntire, D. and Agostini, D. (1970) *Towards Modern Land Policies*, Padua, Institute of Agricultural Economics and Policy.

MacEwen, A. and M. (1982) *National Parks: Conservation or Cosmetics?* London, Allen & Unwin.

McLaughlin, B. P.
(1976) 'Rural settlement planning: a new approach', *Town and Country Planning*, 44, 156-60.
(1981) 'Rural deprivation', *The Planner*, 67, 31-3.

McNee, R. B. (1955) 'Rural development in the Italian South', *Annals of the Association of American Geographers*, 45, 127-51.

Mallet, M. (1978) 'Agriculture et tourisme dans un milieu haut-alpin: un exemple briançonnais', *Etudes Rurales*, 71-2, 111-54.

Manten, A. A. (1975) 'Fifty years of rural landscape planning in the Netherlands', *Landscape Planning*, 2, 197-217.

Mathieu, N.
(1972) 'Le rôle des petites villes en milieu rural', *Bulletin de l'Association de Géographes Français*, 400-1, 287-94.
(1974) 'Propos critiques sur l'urbanisation des campagnes', *Espaces et Sociétés*, 12, 71-89.

Mathieu, N. and Bontron, J. C. (1973) 'Les transformations de l'espace rural: problèmes de méthode', *Etudes Rurales*, 49-50, 137-59.

Matthews, G. V. T. (1971) 'Wetlands in the natural landscape', *Nature in Focus*, 9, 17-20.

Mayhew, A.
(1970) 'Structural reform and the future of West German agriculture', *Geographical Review*, 60, 54-68.

(1971) 'Agrarian reform in West Germany: an assessment of the integrated development project Mooriem', *Transactions, Institute of British Geographers*, 52, 61–76.

(1973) *Rural Settlement and Farming in Germany*, London, Batsford.

Mead, W. R. (1966) 'The study of field boundaries', *Geographische Zeitschrift*, 54, 101–17.

Medici, G. (ed.) (1969) *World Atlas of Agriculture: Europe*, Novara, Istituto Geografico de Agostini.

Meijer, H.

(1975) 'Zuyder Zee: Lake Ijssel', *Bulletin, Information and Documentation Centre for the Geography of the Netherlands*, 1–72.

(1980) 'The region of the great rivers', *Bulletin, Information and Documentation Centre for the Geography of the Netherlands*, 7–33.

Mendras, H.

(1979) *Voyage au Pays de l'Utopie Rustique*, Le Paradou, Actes Sud.

(1980) *La Sagesse et le Désordre, France 1980*, Paris, Gallimard.

Métailié, J. P. (1978) 'Les incendies pastoraux dans les Pyrénées centrales', *Revue Géographique des Pyrénées et du Sud-Ouest*, 49, 517–26.

Meynier, A. (1982) 'Les campagnes des Ségalas et du Levézou', *Annales de Géographie*, 91, 638–9.

Miège, J. (1976) 'Le parc national des îles d'Hyères', *Revue de Géographie de Lyon*, 51, 133–49.

Mignon, C. (1971) 'L'agriculture à temps partiel dans le département du Puy-de-Dôme, *Revue d'Auvergne*, 85, 1–41.

Milhau, J. and Montagne, R. (1961) *L'Agriculture Aujourd'hui et Demain*, Paris, Presses Universitaires de France.

Mills, S. (1983) 'French farming: good for people, good for wildlife', *New Scientist*, 568–71.

Mioni, A. (1978) 'Regional planning and territorial interventions in Italy 1880–1940', *Landscape Planning*, 5, 213–37.

Moore, D. (ed.) (1979) *Disadvantaged Rural Europe: Development Issues and Approaches*, Langholm, Arkleton Trust.

Morandini, R. (1978) 'Survey and analysis for forestry and soil protection in Italy', *Landscape Planning*, 5, 181–92.

Mormont, M. (1983) 'The emergence of rural struggles and their ideological effects', *International Journal of Urban and Regional Research*, 7, 559–75.

Moseley, M. J. (1979) *Accessibility: the Rural Challenge*, London, Methuen.

Mougenot, C. (1982) 'Les mécanismes sociaux de la rurbanisation', *Sociologia Ruralis,* 22, 264-78.

Mrohs, E.
(1982) 'Part-time farming in the Federal Republic of Germany', *GeoJournal,* 6, 327-30.

(1983) 'Zur sozialen Lage der Nebenerwerbslandwirte in der Bundesrepublik Deutschland 1980', *Sociologia Ruralis,* 23, 28-49.

Muller, B. (1973) 'Le vignoble de Novéant-sur-Moselle: histoire d'un déclin', *Mosella,* 3, 1-39.

Munton, R. J. C. (1983) *London's Green Belt,* London, Allen & Unwin.

Natali, L. (1983) 'For the southern regions of the Community: the integrated Mediterranean programmes', *Green Europe,* 197, 1-128.

Naudet, G. (1976) 'France: the regional parks', *Nateuropa,* 26, 12-18.

Naveh, Z. (1982) 'Mediterranean landscape evolution and degradation', *Landscape Planning,* 9, 125-46.

Naylor, E. L. (1982) 'Retirement policy in French agriculture', *Journal of Agricultural Economics,* 33, 25-36.

Newby, H. (1977) 'Farmers' attitudes to conservation', *Countryside Recreation Review,* 2, 23-30.

Newcomb, R. M. (1972) 'Has the past a future in Denmark? The preservation of landscape history within the nature park', *Geoforum,* 9, 61-7.

Niederbacher, A. (1982) *Wine in the European Community,* Luxembourg, European Commission.

Niggermann, J. (1980) 'The development of agrarian structure and cultural landscape', *Bochumer Geographische Arbeiter,* 38, 38-43.

Noirfalise, A. (1979) 'Man's use of the forest', *Nateuropa,* 33, 26-8.

Organization for Economic Co-operation and Development (1977) *Part-time Farming in OECD Countries,* Paris, OECD.

O'Riordan, T. (1983) 'Putting trust in the countryside', in World Conservation Strategy, *The Conservation and Development Programme for the UK,* London, Kogan Page, 171-260.

Orlando, G. and Antonelli, G. (1981) 'Regional policy in EC countries and Community regional policy: a note on problems and perspectives in developing depressed rural regions', *European Review of Agricultural Economics,* 8, 213-46.

Ozenda, P. (1979) 'Vegetation map of the Council of Europe member states', *Council of Europe Nature and Environment Series,* 16.

Pahl, R. E.
(1965) 'Class and community in English commuter villages',

Sociologia Ruralis, 5, 5–23.

(1975) *Whose City?*, Harmondsworth, Penguin.

Paillat, P. and Parent, A. (1980) *Le Vieillissement de la Campagne Francaise*, Paris, Institut National d'Etudes Démographiques.

Patella, L. V. (1981) 'Population redistribution in the northern Apennines (Italy) in recent years', in J. W. Webb (ed.) *Policies of Population Redistribution*, Oulu, Geographical Society of Northern Finland.

Patmore, J. A. (1970) *Land and Leisure*, Newton Abbot, David & Charles.

Pelapsis, A. A. and Thompson, K. (1960) 'Agriculture in a restrictive environment: the case of Greece', *Economic Geography*, 36, 145–57.

Peper, B. (1969) 'Agricultural policy and social policy', *Sociologia Ruralis*, 9, 221–34.

Peterken, G. F. and Harding, P. T. (1975) 'Woodland conservation in eastern England', *Biological Conservation*, 8, 278–98.

Petit, M. (1981) 'Agriculture and regional development in Europe', *European Review of Agricultural Economics*, 8, 137–53.

Phillips, D. R. and Williams, A. M.

(1982a) 'Local authority housing and accessibility', *Transactions, Institute of British Geographers*, 7, 304–20.

(1982b) *Rural Housing and the Public Sector*, Farnborough, Gower Press.

Pichol, M. (1978) 'Territoire à prendre, territoire à défendre: le Larzac', *Hérodote*, 10, 91–112.

Pickup, C. H. and Fairclough, K. (1982) 'Not just for the birds: ecological impact of the IDP in the Uists', *Ecos*, 3, 24–9.

Picon, B. (1978) 'Mécanismes sociaux de transformation d'un éco-système fragile: la Camargue', *Etudes Rurales*, 71–2, 219–29.

Pieroni, O. (1982) 'Positive aspects of part-time farming in the development of a professional agriculture, remarks on the Italian situation', *GeoJournal*, 6, 331–6.

Pilleboue, J. (1972) 'Le nord du Causse du Larzac, une renaissance menacée', *Revue Géographique des Pyrénées et du Sud-Ouest*, 43, 453–68.

Pinchemel, P. (1957) *Structures Sociales et Dépopulation Rurale de la Plaine Picarde de 1836 à 1936*, Paris, Armand Colin.

Pisani, E. (1979) *Agriculture, Environnement et Vie Rurale*, Paris, GREP.

Pitié, J. (1969) 'Pour une géographie de l'inconfort des maisons rurales', *Norois*, 16, 147–76.

Pitte, J-R. (1983) *Histoire du Paysage Français*, 2 vols, Paris, Tallandier.

Planck, U. (1977) 'République fédérale d'Allemagne', *Futuribles*, 107–42.

Poore, D. and Ambroes, P. G. (1980) *Nature Conservation in Northern and Western Europe*, Gland, United Nations Environment Programme.

Pratschke, J. L. (1981) 'Rural and farm dwellings in the European Community', *Irish Journal of Agricultural Economics and Rural Sociology*, 8, 191–211.

Préau, P. (1976) 'Le parc national de la Vanoise et l'aménagement de la montagne', *Revue de Géographie de Lyon*, 51, 123–32.

Raffin, J. P. and Leufeuvre, J. C. (1982) 'Chasse et conservation de la faune sauvage en France', *Biological Conservation*, 23, 217–41.

Rainelli, P. and Bonnieux, F. (1978) 'Situation et évolution structurelle et socio-économique des régions agricoles de la Communauté, *Informations sur l'Agriculture*, 52 and 53.

Rambaud, P.

(1967) 'Tourisme et urbanisation des campagnes', *Sociologia Ruralis*, 7, 311–34.

(1969) *Société Rurale et Urbanisation*, Paris, Seuil.

(1980) 'Tourisme et village: un débat de société', *Sociologia Ruralis*, 20, 232–49.

Reitel, F. (1973) 'Les causes du déclin du vignoble mosellan', *Mosella*, 3, 40–71.

Renzoni, A. (1975) 'Hunting regulations in Italy', *Biological Conservation*, 8, 185–8.

Rhun, P. (1977) 'Destruction d'un paysage: protestations paysannes et réflexions théoriques', *Hérodote*, 7, 52–70.

Richez, G.

(1973) 'Les parcs naturels dans le sud-est de la France', *Méditerranée*, 1, 81–101 and 119–35.

(1983) 'Le parc naturel régional de la Corse', *Méditerranée*, 47, 35–44.

Rigole, A. (1972) 'Le rapport actuel de l'homme et de la forêt en Basse-Provence et Côte d'Azur', *Acta Geographica*, 12, 179–98.

Roberts, S. and Randolph, W. G. (1983) 'Beyond decentralization: the evolution of population redistribution in England and Wales 1961–1981', *Geoforum*, 14, 75–102.

Roderkerk, E. C. M. (1974) 'Recreation and nature conservation in the Kennermerduinen national park', *Nature in Focus*, 18, 7–10.

Rogers, A. W.
    (ed.) (1978) *Urban Growth, Farmland Losses and Planning*, Wye
    College.
    (1981) 'Housing in the national parks', *Town and Country Planning*,
    50, 193–5.
Rogers, R. (1979) *Schools under Threat: a Handbook on Closures*,
    London, Advisory Centre for Education.
Rose, C. (1982) 'Of global concern', *Ecos*, 3, 20–1.
Rosenberg, H. (1973) 'Workers in French Alpine tourism: whose
    development?', *Studies in European Society*, 1, 21–38.
Roudié, P. (1978) 'Le rôle de techniques dans l'évolution des paysages
    viticoles d'Europe occidentale', *Geographia Polonica*, 38, 253–6.
Roux, J. M. (1977) 'Dévitalisation d'un pays', *Economie Rurale*, 117,
    17–27.
Ruggieri, M.
    (1972) 'Modificazioni degli abitati abruzzesi', *Bollettino della Società
    Geografica Italiana*, 10(1), 487–505.
    (1976) 'I terreni abbandonati: nuova componente del paesaggio',
    Bolletino della Società Geografica Italiana, 10(5), 441–64.
Ruppert, K. (1980) 'Basic tendencies of space structure influenced by
    leisure activities', *Bochumer Geographische Arbeiter*, 38, 44–53.
Sachs, I.
    (1978) 'Ecodéveloppement: une approache de planification',
    Economie Rurale, 124, 16–22.
    (1980) *Stratégies de l'Ecodéveloppement*, Paris, Editions Ouvrières.
Santos, M. (1978) 'De la société au paysage', *Hérodote*, 9, 66–73.
Saurin, J. P. (1980) 'The compatibility of conifer afforestation with the
    landscape of the Monts d'Arrée region', *Landscape Planning*, 7,
    295–311.
Scheifele, M. (1979) 'Europe's forests', *Nateuropa*, 33, 24–5.
Schéma Général d'Aménagement de la France (1981) 'La France rurale:
    images et perspectives', *Travaux et Recherches de Prospective, Docu-
    mentation Française*, 81, 1–164.
Schwarzweller, H. K. (1971) 'Tractorization of agriculture: the social
    history of a German village', *Sociologia Ruralis*, 11, 127–39.
Scottish Consumer Council (1982) *Consumer Problems in Rural Areas*,
    Glasgow.
Sermonti, E. (1968) 'Agriculture in areas of urban expansion: an
    Italian study', *Journal of the Town Planning Institute*, 54, 15–17.
Shaw, J. M. (1979) *Rural Deprivation and Planning*, Norwich,
    Geobooks.

Sheail, J. (1975) 'The concept of national parks in Great Britain 1900–1950', *Transactions, Institute of British Geographers*, 66, 41–56.

Shoard, M.
(1980) *The Theft of the Countryside*, London, Temple Smith.
(1981) 'Why landscapes are harder to protect than buildings', in D. Lowenthal and M. Binney (eds) *Our Past before Us*, London, Temple Smith, 83–108.

Sindt, M. (1971) 'Le vignoble luxembourgeois', *Mosella*, 1, 2–23.

Sperling, W. (1966) 'Modern evolution in the rural landscape of the Federal Republic of Germany', *Acta Geologica et Geographica Universitatis Comenianae*, 6, 175–91.

Taylor, C. and Emerson, D. (1981) *Rural Post Offices: Retaining a Vital Service*, London, Bedford Square Press.

Tellegen, E. (1981) 'The environmental movement in the Netherlands', in T. O'Riordan and R. K. Turner (eds) *Progress in Resource Management and Environmental Planning 3*, 1–31.

Terrasson, F. and Tendron, G. (1981) 'The case for hedgerows', *The Ecologist*, 11, 210–21.

Thénoz, M. (1981) 'La pratique touristique estivale et son impact dans un espace protégé', *Revue de Géographie de Lyon*, 56, 275–302.

Thiede, G. (1976) 'Capacité de l'agriculture européenne et son avenir', *Economie Rurale*, 116, 47–52.

Thirgood, J. V. (1981) *Man and the Mediterranean Forest*, London, Academic Press.

Thissen, F. (1978) 'Second homes in the Netherlands', *Tijdschrift voor Economische en Sociale Geografie*, 69, 322–32.

Thorpe, H. (1975) 'The homely allotment: from rural dole to urban amenity', *Geography*, 60, 169–83.

Tirone, L. (1975) 'Mutations récentes du vignoble italien', *Méditerranée*, 23, 59–80.

Toepfen, A. (1981) 'A priceless heritage', *Nateuropa*, 38, 21–2.

Tovias, A. (1979) 'EEC enlargement: the southern neighbours', *Sussex European Papers*, 5, 1–80.

Tracy, M.
(1982a) *Agriculture in Western Europe: Challenge and Response 1880–1980*, London, Granada.
(1982b) 'People and policies in rural development: institutional problems in the formulation and implementation of rural development policies in the European Community', *The Arkleton Trust Seminar Papers*, 1–56.

Truchis, M. (1978) 'L'amélioration des transports publics en zone rurale faiblement peuplée', *Economie Rurale*, 125, 55-8.

Tsoukalis, L. (1981) *The European Community and its Mediterranean Enlargement*, London, Allen & Unwin.

Tuppen, J. N. (1983) *The Economic Geography of France*, London, Croom Helm.

Turton, B. J. (1970) 'The western Po basin in Italy: a study in industrial expansion and the journey to work', *Town Planning Review*, 41, 357-71.

Van der Haegen, H. (1982) 'West European settlement systems', *Acta Lovaniensia*, 22, 1-370.

Vanlaer, J. (1979) 'Les villages de vacances dans la région wallonne', *Revue Belge de Géographie*, 103, 151-209.

Van Lier, H. N. and Steiner, F. R. (1982) 'A review of the Zuiderzee reclamation works', *Landscape Planning*, 9, 35-59.

Van Lier, H. N. and Taylor, P. D. (1982) 'The new Dutch landscape', *Landscape Architecture*, 72, 66-71.

Vincent, J. A. (1980) 'The political economy of Alpine development: tourism or agriculture in Saint Maurice', *Sociologia Ruralis*, 20, 250-71.

Vincent, M. (1982) 'France' in Fondation Rurale de la Wallonie, *Initiative Economique en Milieu Rural*, Arlon, 23-32.

Von Kürten, W. (1980) 'Landscape preservation and landscape management', *Bochumer Geographische Arbeiter*, 38, 61-7.

Wagstaff, J. M.
   (1965) 'Traditional houses in modern Greece', *Geography*, 50, 58-64.
   (1968) 'Rural migration in Greece', *Geography*, 53, 175-9.

Walker, A. (1978) *Rural Poverty*, London, Child Poverty Action Group.

Wallace, D. B. (1981) 'Rural policy', *Town Planning Review*, 52, 215-22.

Warnes, A. M. (ed.) (1982) *Geographical Perspectives on the Elderly*, Chichester, Wiley.

Warren, A. and Goldsmith, F. B. (eds) (1983) *Conservation in Perspective*, Chichester, Wiley.

Weinschenck, G. and Kemper, J. (1981) 'Agricultural policies and their regional impact in Western Europe', *European Review of Agricultural Economics*, 8, 251-81.

Westmacott, R. and Worthington, T. (1974) *New Agricultural Landscapes*, London, Countryside Commission.

White, P. E. (1976) 'Tourism and economic development in the rural environment', in R. Lee and P. E. Ogden (eds) *Economy and Society in the EEC*, Farnborough, Saxon House, 150-9.

Wibberley, G. P.

(1960) 'Changes in the structure and functions of the rural community', *Sociologia Ruralis*, 1, 118-27.

(1961-2) 'Agriculture and land-use planning', *Town Planning Review*, 32, 77-94.

(1974) 'The proper use of Britain's rural land', *Journal of the Royal Town Planning Institute*, 60, 79-82.

(1975) 'Changes in the human social environment brought about by resource problems', in University of Guelph, *Agriculture in the Whirlpool of Change*, Guelph, 84-96.

(1981) 'Strong agricultures but weak rural economies: the undue emphasis on agriculture in European rural development', *European Review of Agricultural Economics*, 8, 155-70.

Wild, T.

(1981) *West Germany: a Geography of its People*, London, Longman.

(ed.) (1983) *Urban and Rural Change in West Germany*, London, Croom Helm.

Wood, C. (1982) 'Implications of the European EIA initiative', *Ecos*, 3, 25-9.

Woolmore, R. G. (1971) 'A European national park: the Gran Paradiso', *Recreation News Supplement*, 3, 13-16.

Zetter, J. A. (1970) 'Het Nationale Park de Kennermerduinen', *Recreation News Supplement*, 1, 2-5.

Select bibliography

Best, R. H. (1981) *Land Use and Living Space*, London, Methuen.

Béteille, R. (1981) *La France du Vide*, Paris, Librairies Techniques.

Body, R. (1982) *Agriculture: the Triumph and the Shame*, London, Temple Smith.

Bowers, J. K. and Cheshire, P. (1983) *Agriculture, the Countryside and Land Use*, London, Methuen.

Bryant, C. R., Russwurm, L. H. and McLellan, A. G. (1983) *The City's Countryside: Land and its Management in the Rural-urban Fringe*, London, Longman.

Calmes, R. (1978) *L'Espace Rural Français*, Paris, Masson.

Cloke, P.
(1979) *Key Settlements in Rural Areas*, London, Methuen.
(1983) *An Introduction to Rural Settlement Planning*, London, Methuen.

Coppock, J. T. (ed.) (1977) *Second Homes: Curse or Blessing?*, Oxford, Pergamon.

Davidson, J. and Lloyd, R. (eds) (1977) *Conservation and Agriculture*, Chichester, Wiley.

Duffey, E. (1982) *National Parks and Reserves of Western Europe*, London, Macdonald.

Franklin, S. H.
(1969) *The European Peasantry: the Final Phase*, London, Methuen.
(1971) *Rural Societies*, London, Macmillan.

Green, B. H. (1980) *Countryside Conservation*, London, Allen & Unwin.

Hall, P. G. and Hay, D. (1980) *Growth Centres in the European Urban System*, London, Heinemann.

Hodge, I. and Whitby, M. (1981) *Rural Employment: Trends, Options, Choices*, London, Methuen.

Johnson, J. H. (ed.) (1976) *Suburban Growth*, Chichester, Wiley.

MacEwen, A. and M. (1982) *National Parks: Conservation or Cosmetics?*, London, Allen & Unwin.

Moseley, M. J. (1979) *Accessibility: the Rural Challenge*, London, Methuen.

O'Riordan, T. (1983) 'Putting trust in the countryside', in World Conservation Strategy, *The Conservation and Development Programme for the UK*, London, Kogan Page, 171–260.

Pye-Smith, C. and Rose, C. (1984) *Crisis and Conservation: Conflict in the British Countryside*, Harmondsworth, Penguin.

Rogers, A. W. (ed.) (1978) *Urban Growth, Farmland Losses and Planning*, Wye College.

Shaw, J. M. (1979) *Rural Deprivation and Planning*, Norwich, Geobooks.

Shoard, M. (1980) *The Theft of the Countryside*, London, Temple Smith.

Tracy, M. (1982) *Agriculture in Western Europe: Challenge and Response, 1880–1980*, London, Granada.

Walker, A. (1978) *Rural Poverty*, London, Child Poverty Action Group.

Warren, A. and Goldsmith, F. B. (eds) (1983) *Conservation in Perspective*, Chichester, Wiley.

Westmacott, R. and Worthington, T. (1974) *New Agricultural Landscapes*, London, Countryside Commission.

Williams, A. (ed.) (1984) *Southern Europe Transformed*, London, Harper & Row.

Index

For Product Safety Concerns and Information please contact our EU
representative GPSR@taylorandfrancis.com
Taylor & Francis Verlag GmbH, Kaufingerstraße 24, 80331 München, Germany

www.ingramcontent.com/pod-product-compliance
Ingram Content Group UK Ltd.
Pitfield, Milton Keynes, MK11 3LW, UK
UKHW020935180425
457613UK00019B/401